HOW TO BUILD A
BRAIN

HOW TO BUILD A
BRAIN

AND 34 OTHER REALLY INTERESTING USES OF MATHEMATICS

RICHARD ELWES

Quercus

Contents

$$x^3 + x = 11$$

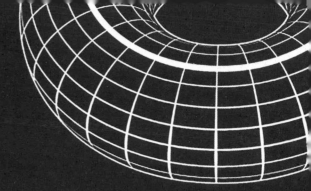

Introduction

Mathematics without the boring bits? Surely that must be a joke? Everyone knows there is no more unwelcoming and technically forbidding subject than mathematics, couched as it is in its own language of incomprehensible diagrams, arcane symbols and obscure jargon.

Alas, that view is all too common, reinforced for many people by painful recollections of tedious homework, or bruising encounters with tough exams. Beyond chores such as adding up bills, or calculating times and dates, most people are happy to leave mathematics to the experts. Sure, it's great that computer programmers, engineers and physicists find it useful for what they do; as long as they don't bother us with the details, we can happily enjoy the fruits of their labours.

The stereotype also attaches itself to the practitioners of mathematics, long perceived as brains on stalks, somehow devoid of normal human attributes. Certainly, you'd want one of them in your pub quiz team. But would you want their conversation at the bar afterwards, or – heaven forbid – at your dinner party?

So it's time to slay some ghosts and kill some prejudices. Let's admit that any field of human endeavour worth investigating eventually gets to a level of technicality that becomes challenging, and it's fair to say that many people hit that ceiling earlier in mathematics before they do in other, perhaps more verbal, subjects. But before that level is reached, there is a whole accessible world of mathematics that can astonish with its diversity, captivate with its mystery, and enthral with its beauty.

So vast is today's mathematics that it encompasses almost any description – and its opposite – that one might care to try. For a start, mathematics is both ancient and modern, built by traditional methods over many centuries, and yet its face always set firmly to the future. In the 'cradle of civilization' of the Middle East, Babylonian mathematicians were developing sophisticated systems for counting. Later, Greek thinkers such as Plato and Euclid made stunning discoveries in geometry and number theory. From this grand lineage has sprung an irrepressible modernity, for mathematics also lies at the heart of the latest human advances in science and technology, from computer-code to comprehending the cosmos. Mathematics covers the microscopically small, the infinitely large, and everything between. It is vital to those who spend their lives investigating tiniest sub-atomic particles, and it fuels the latest hypotheses about a still-expanding universe. Mathematics is the discovery of what exists, but it is also creation anew. It provides the principles to describe and measure the natural world. Yet it is also an extraordinary collective act of creativity, spanning thousands of years. To put it another way, mathematicians must be as rigorous and analytical as they are passionate and imaginative. They love nothing more than the rock-solid certainty of a proof, but are unafraid to contemplate the most outrageous 'What if?'

Mathematics is as much about the unknown as it is the known. The old, enduring mysteries continue to fascinate and beguile. And while a proven theorem brings the immense satisfaction of a conundrum finally conquered and understood, at the same time each new discovery prompts ten new questions, as our collective understanding probes ever greater depths.

Today, the Clay Institute is offering lucrative prizes to the solvers of several outstanding questions, each of which could have profound implications for our lives. If one puzzle, the succinctly named 'P = NP?', should be resolved, then the security of the world's computer networks may lie in tatters. Mathematics matters.

Mathematics is impeccably logical – until it hits a paradox. On the one hand, it seems to revel in tidy conclusions, a jigsaw-puzzle world in which every piece slots into place. If $a + b = 5$, and we know $a = 2$, then we confidently state that b is 3. But it also confounds expectations. Once you have reached infinity, can you count any further? Common sense says 'no', but in 19th century Germany, Georg Cantor dared to say 'yes'. Mathematics demands proof, and it was Cantor, and not common-sense, who prevailed.

Mathematics can be wild and unruly, but also serene and delicate. Chaos theory predicts that tiny differences in circumstance can produce dramatically differing outcomes. But mathematical principles also underlie aesthetics, whether that means symmetry, proportion or regularity, and whether we are talking about a handsome face, a pretty wallpaper pattern, or a fugue by J.S. Bach. For mathematical connoisseurs, a concise formula (such as that of the circle) is itself suffused with a minimalistic elegance, providing a satisfying answer to a natural question.

One could go on. But perhaps the mathematics should now speak for itself. This book – a guide to 35 landmarks of the mathematical world – is for the mathematical sightseer. I hope you will venture out and see the sights, learn a little, be surprised - and perhaps occasionally amazed. Above all, enjoy the journey. Bon voyage.

1 How to solve every equation there has ever been

- Counting and natural numbers
- Negative numbers and trade
- Rational and irrational numbers
- Real and imaginary numbers
- Complex numbers and the fundamental theorem of algebra

$$x^3 + x + x = 11$$

Since the first hunters counted their haul at the end of the day, humans have employed numbers as vital tools for understanding the world. In the intervening years, just as human civilization has developed beyond recognition so the same is true of our numbers. Today's numbers are flashy, modern constructions, whose development coincided with critical moments of history. At each stage, the notion of numbers had to be expanded to include controversial new entities.

What are numbers good for?

The first thing numbers are used for is counting: three trees, two children, 100 enemy warriors. As soon as we have this idea, we can use numbers to solve problems. For example, suppose my family of four each need two jugs of water to drink each day, and I need another three jugs to water my vegetable garden. How many jugs do I need altogether? In numerical terms, the problem is $(4 \times 2) + 3 = ?$

Today's mathematicians prefer to use letters such as x and y to represent unknown numbers. So the problem would be rewritten as $(4 \times 2) + 3 = x$.

This is an *equation*. To solve it, all we have to do is perform the calculations, to get the solution: $x = 11$. But not all equations are so easy to solve.

Rational numbers: divide and conquer

Suppose I have two children but my day's hunting has produced just one rabbit. How can I share the food between them? There is no way to do it, with each child getting a whole number of rabbits. Such questions heralded the introduction of a new type of number, known to most people as a *fraction*; $\frac{1}{2}$ is just another way of writing $1 \div 2$. Mathematicians prefer to call fractions *rational numbers*. (This comes from the fact that they are *ratios* of whole numbers; it is nothing to do with being sensible.)

Negative numbers and the dawn of trade

Critical to the development of human society was the beginning of trade, which allowed people to specialize in a particular line of work. One person could hunt for meat, another grow vegetables, and a third make shoes. Then they could trade to ensure that everyone had meat, vegetables and shoes.

A day's trade can end in profit, in loss or break even. To measure profit and loss on a single scale, early mathematicians came up with a new number

The natural numbers   The integers

system today called the *integers*. The system of integers combines the positive whole numbers with new *negative numbers*: $-1, -2, -3, -4, -5, -6, \ldots$

From the perspective of solving equations, this new system has a great advantage. Any equation that involves only addition and subtraction can now be solved. For example, before the dawn of negative numbers, the equation $x + 4 = 3$ would have been considered meaningless. Interpreted in terms of trade, however, it makes perfect sense. If I end the day three apples in credit having earnt four, then in the morning I must have been in debt by one apple. So the solution is $x = -1$.

● Minus and minus makes plus

Once we have got negative numbers, we need to know how to manipulate them with the usual rules: addition, subtraction, multiplication and division. Addition is fairly easy: $(-3) + (-5) = (-8)$. Going the other way, $(-3) - (-5) = 2$. The best way to see this is by using an infinite number line: adding negative numbers involves moving left, as does subtracting positive numbers. Adding positive numbers means moving right, as does subtracting negative numbers.

The trickiest aspect is multiplication. The key rule is this: when you multiply two positive numbers, you get a positive answer. One negative and one positive number produce a negative answer. Most confusingly, two negative numbers give a positive result.

Why should this be? Suppose I make 2 units of profit every day. In three days' time I will be 6 units up on today ($2 \times 3 = 6$). Three days ago, however (that is to say, in -3 days' time), I was 6 units down on today ($2 \times (-3) = (-6)$).

Now, what if I lose 2 units every day? Then in three days' time I will be 6 units down on today (that is, $(-2) \times 3 = (-6)$). Three days ago, however, I was 6 units up on today ($(-2) \times (-3) = 6$).

● The limits of rationality

The system of rational numbers (positive, negative and zero) can describe many aspects of the world. Mathematically it works well too, as we can add, subtract, multiply and divide within this system, never running out of numbers. Perhaps this system represents all we will ever need? As it turned out, the history of numbers had much further to run.

Unfortunately, the rational numbers are inadequate for even the most elementary geometry. If we start with a square, 1 cm by 1 cm, what is the length of its diagonal? Call this number c. According to Pythagoras' theorem, it must satisfy $1 + 1 = c \times c$ (see *How to become a celebrity mathematician*). In other words, we need to solve the equation $c^2 = 2$. The technical term is the *square root* of 2, written $c = \sqrt{2}$. What is this number?

An approximate answer is $1\frac{2}{5}$. However, the disconcerting truth is that there is no fraction that satisfies this requirement exactly. In modern terms $\sqrt{2}$ is what is known as an *irrational number*, meaning that it can never be expressed exactly as a fraction. Another famous example is π.

The ghosts between numbers

Elementary geometry shows that the rational numbers are inadequate for a great deal of mathematics. If we arrange the rational numbers in a long line, there are holes in between them. This is not obvious at first sight, because rational numbers can be very close together, indeed as close together as you like. Nevertheless, there is a hole at $\sqrt{2}$ and there is another at π. Unexpectedly, there are not just one or two holes; the whole line is riddled with them. This is where the irrational numbers live, and most are far less easy to describe than $\sqrt{2}$ or π.

Although the existence of irrational numbers had been known for thousands of years, it was not until the 19th century that the right way was found to expand the rational numbers to a larger system, filling in the gaps. This is called the system of *real numbers*. It is a flashy, powerful, modern framework. But the evolution of numbers had still not yet run its course.

Solving equations

From a mathematician's perspective, the real numbers represent a huge step forward. A whole spectrum of equations that were previously insoluble suddenly have solutions. The number $\sqrt{2}$ solves the equation $x^2 = 2$. Similarly we can now solve the equations $x^2 = 3$, $x^2 = 5$, $x^2 = 6$, as well as more complicated equations such as $x^3 + x = 11$, and far beyond.

However, not every equation has a solution, even in the real numbers. When you multiply a positive number by itself, you get a positive answer. Similarly, when you multiply a negative number by itself, you also get a positive

'The irrationality of a thing is no argument against its existence, rather a condition of it.'

FRIEDRICH NIETZSCHE

The real numbers form a line with no holes

answer. That means there is no real number that, when multiplied by itself, gives a negative answer. In other words, the equation $x \times x = -1$ (or $x^2 = -1$) has no solutions in the real numbers. There is no $\sqrt{-1}$ in the real numbers.

● Imaginary numbers

Both negative numbers and irrational numbers were highly controversial when first introduced. But these steps were nothing compared with the next dramatic expansion of the concept of number, to that of *imaginary numbers*.

Among the first people to contemplate imaginary numbers were Italian scholars of the 16th century. In this period, solving ever more complicated equations assumed a role of intellectual gladiatorial combat. The greatest mathematicians of the age would meet for public bouts of equation-solving. These warriors did not embrace imaginary numbers as some sort of philosophical ideal. Rather, they came to recognize the computational power they brought. Often, objects such as $\sqrt{-1}$ would appear in their working. Most would abandon the calculation at this stage, believing that it had degenerated into nonsense. But some decided to push on, not worrying too much about what $\sqrt{-1}$ meant. Then, whenever they found $\sqrt{-1} \times \sqrt{-1}$ in their work, they could replace it with -1 and continue. Those who took this bold step were well rewarded. At the end of their working they would often find the 'imaginary' terms had cancelled each other out, leaving them with a good, honest, real solution to their equation.

● Numbers become 2-dimensional

Once the utility of imaginary numbers had been established, it was only a matter of time before the ultimate step was taken, of formally expanding the number system again to include them. To achieve this, the quantity $\sqrt{-1}$ was given the name of i.

The system of *complex numbers* is built from the real numbers together with the imaginary numbers. More precisely, a complex number is the result of adding a real number (such as 4) to an imaginary number such as $3 \times i$ (or $3i$), in this case giving $4 + 3i$. This was done with total mathematical rigour, so at this stage imaginary numbers stopped being purely imaginary and could be added and multiplied in the same way as any other numbers:

$$(4 + 3i) + (2 - i) = 6 + 2i$$

There is a nice pictorial representation of the complex numbers, called the *Argand diagram* (opposite). In this, the real numbers form a horizontal axis, and the imaginary numbers a vertical axis. Every point on the plane can now be represented by giving its real coordinate and its imaginary coordinate. Just as adding or subtracting 1 corresponds to moving right or left along the number line, now adding or subtracting i corresponds to moving up or down. Multiplying by i corresponds to rotating by 90° anticlockwise.

Mathematics at the end of history

The complex numbers allow many more equations to be solved than were possible in the real numbers. In fact, the very definition of the new system is that the equation $x^2 = -1$ has a solution. An unwelcome thought occurs at this stage: are we going to have to expand the system yet again to incorporate solutions to the next level of equations, such as $x^2 = i$?

At the start of the 19th century, Jean Robert Argand produced the first proof of a momentous fact: the complex numbers are now enough to solve every possible equation. More formally, any polynomial you can write down using complex numbers has a solution in the complex numbers. Shortly afterwards, Carl Friedrich Gauss produced another proof of the same result, which came to be known as the *fundamental theorem of algebra*. After thousands of years, the painful process of extending number systems to solve new equations was finally at an end.

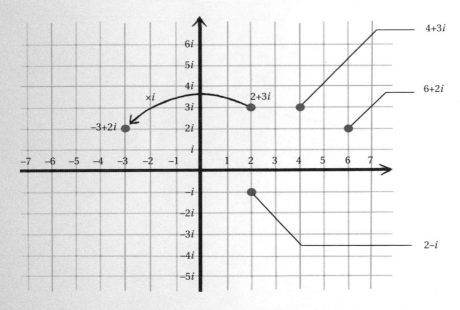

2 How to become a celebrity mathematician

- The beginnings of geometry
- Triangle and Pythagoras' theorem
- Euler bricks and perfect cuboids
- Pierre de Fermat's last theorem

A right-angled triangle is simply a triangle that contains a right angle. Perhaps the first question that every geometry student asks is: what is so interesting about that? To start with, the humble right-angled triangle is the setting for one of the most ancient and illustrious of all mathematical results, Pythagoras' theorem. Although thousands of years old, this theorem is directly connected to some of the deepest questions in modern mathematics, including Fermat's last theorem, finally resolved in 1995, and the search for the elusive perfect cuboid, which continues to defy today's mathematicians.

There is a practical answer too, in that right-angled triangles are the simplest shapes built from straight lines. More complex shapes can then be built from them: squares and rectangles, and then higher-dimensional shapes such as cubes and cuboids.

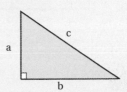

A right-angled triangle

The beginnings of geometry

Pythagoras' theorem describes a relationship between the sides of a right-angled triangle. The longest side is always the one opposite the right angle, and is known as the *hypotenuse*. From the two shorter sides, Pythagoras described a way to work out the length of the hypotenuse.

We take each side of the triangle in turn, measure it, and then *square* it, meaning multiply it by itself. So if the first side has length 3, its square is $3 \times 3 = 9$. This is written as $3^2 = 9$.

When we do this to each side of the triangle, Pythagoras' theorem asserts that the squares of the two shorter sides will add up to the square of the longest side. In algebraic terms, if a and b are the shorter sides, and c is the longest then, $a \times a + b \times b = c \times c$, or more concisely $a^2 + b^2 = c^2$. Geometrically, this means that if you build a square on each side of the triangle the squares on the two shortest sides have a combined area equal to the square on the hypotenuse.

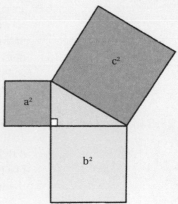

Pythagoras' theorem

Pythagoras: the mystic mathematician

Living around 500 BC, Pythagoras is said to have travelled extensively, to India, Judaea and Egypt, among other places, in search of mathematical wisdom. He was not only a mathematician, but a philosopher and something of a religious guru, the leader of a group called the *mathematikoi*, whom he insisted should foreswear all personal possessions and survive on a strictly vegetarian diet. It is from the *mathematikoi* that we get the word *mathematics*.

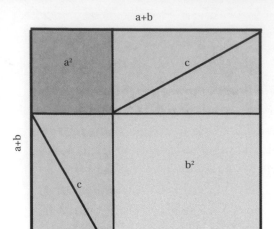

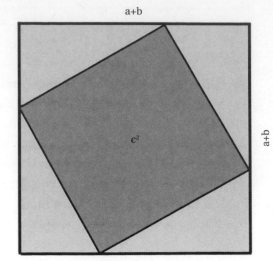

As for the theorem that bears his name, it is true that Pythagoras and the mathematikoi spent much time investigating right-angled triangles, and were fascinated by this theorem and its consequences. It is possible too that they found the first proof of this theorem, thereby showing that it must hold in every possible right-angled triangle. One proof is illustrated above. Draw a square with each side $a + b$ units long. Then it can be divided into four copies of the original triangle, along with one square of area a^2 and one square of area b^2. Dividing it up a different way produces four triangles, together with one square of area c^2. It must be, therefore, that $a^2 + b^2 = c^2$.

● Follow the Euler brick road

However interested in this fact the mathematikoi may have been, knowledge of this result predates Pythagoras by many centuries. Unfortunately we do not know the name of the Babylonian or Indian geometers who discovered it, so Pythagoras can continue to take the plaudits.

These early appearances of the Pythagorean theorem do not state it explicitly, but contain tell-tale triples of numbers such as 3, 4, 5 and 5, 12, 13. To the trained eye, these are instantly recognizable as 'Pythagorean triples'. A Pythagorean triple comprises three numbers that correspond to the sides of a right-angled triangle. It is easy to check that they satisfy Pythagoras' theorem, for example $3^2 + 4^2 = 9 + 16 = 25 = 5^2$.

There is something more to notice about these Pythagorean triples. Not only are they the lengths of the sides of a right-angled triangle, but they are also whole numbers. This is highly unusual: most triangles do not have whole numbers for all three lengths. For example, if you draw two lines of length 1 cm, at right angles, and connect their ends, the third side has length $\sqrt{2}$ cm. This is not a whole number; in fact it cannot even be written as a fraction, but is *irrational* (see *How to square a circle*). This is the usual scenario. So Pythagorean triples are of enduring significance: they represent the few occasions when the simplest geometrical shape, the right-angled triangle, meets the simplest type of number, the whole number. The first two examples have sides (3, 4, 5) and (5, 12, 13), then (8, 15, 17) and (7, 24, 25). Any multiple of these will also be a Pythagorean triple (for example (6, 8, 10) is twice (3, 4, 5)).

One of the reasons why right-angled triangles are so useful is that other shapes are built from them. The simplest example is the rectangle. If you cut a rectangle in two diagonally, then you are left with two identical right-angled triangles. So if a particular rectangle has sides of length 5 cm and 12 cm, then we can use Pythagoras' theorem to determine the length of the diagonals, in this case 13 cm. Pythagorean triples therefore produce rectangles whose sides and diagonals are all whole numbers.

The same thing holds when we step into three dimensions. Just as a rectangle is a stretched square, a *cuboid* is a stretched cube. It has six faces, all rectangular, with opposite faces identical. Of course, once we know the lengths of the edges, Pythagoras' theorem will give us the lengths of the diagonals of each face.

The great Swiss mathematician Leonhard Euler was interested in finding cuboids where all the edges are whole numbers, as are the diagonals on each face. A cuboid like this is called an *Euler brick*, a 3-dimensional analogue of a Pythagorean triple. The difference is that Euler bricks are much harder to find. The first Euler brick was discovered not by Euler but by his contemporary Paul Halcke in 1719. It has edges of length 44, 117 and 240 and its faces have diagonals of 125, 244 and 267. We now know that the list of Euler bricks, like that of Pythagorean triples, is endless: there are infinitely many Euler bricks.

The quest for the perfect cuboid

The definition of an Euler brick is that the edges are all whole numbers, as are the lengths of the diagonals on each face. An obvious extra requirement is that the *body diagonals* should also be given by a whole number. This is now the definition of a *perfect cuboid*. Unfortunately, none has ever been found. Mathematical celebrity awaits anyone able to mould a brick with all its dimensions given by whole numbers (lengths, face diagonals and body diagonals), or

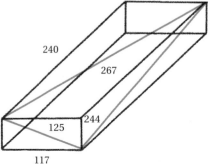

alternatively to show that no such shape can ever exist. The search will have to be extensive, however, because if a perfect cuboid does exist at least one of its edges must be more than 10,000,000,000 units long. All smaller possibilities have been ruled out.

Although a perfect cuboid has never been found, in 2009 Jorge Sawyer and Clifford Reiter did discover a *perfect parallelepiped*. A *parallelepiped* resembles a cuboid that has been sat on. Instead of having rectangular faces, it has faces that are *parallelograms*, shapes made from two sets of parallel lines (like a rectangle), but not meeting at right angles. The smallest of Sawyer and Reiter's perfect parallelepipeds has edges of length 271, 106, 103, face diagonals of length 101 and 183, 266 and 312, 255 and 323, and body diagonals of length 374, 300, 278 and 272.

From geometry to algebra: Fermat's last theorem

Although it has its roots in geometry, the problem of Euler's bricks is part of a different mathematical discipline, called *number theory*. Around AD 250, the mathematician Diophantus wrote his celebrated work *Arithmetica*. In it he considered Pythagoras' formula $a^2 + b^2 = c^2$, as well as a variety of other equations. Once we have this equation, we can throw away the picture of the triangle. But, in each case, Diophantus was not interested in general solutions to the equations; he only cared about finding *whole numbers* that satisfy it. These are now called *Diophantine problems*.

The problem of the perfect cuboid is one of the most famous of all Diophantine problems. Another is *Fermat's last theorem*. Again, this has its origins in Pythagoras' theorem. The Pythagorean triple 3, 4, 5 is a solution to Pythagoras' equation: $a^2 + b^2 = c^2$ (because $3^2 + 4^2 = 5^2$). Diophantus thought of this as splitting one square c^2 into two smaller squares, a^2 and b^2.

Just as Leonhard Euler's cuboids later lifted this into three dimensions geometrically, so Pierre de Fermat generalized this question in a different direction, using algebra. By profession, Fermat was a lawyer and French government official, but his passion was for mathematics. Through his own research and in correspondence with other mathematicians of the age, he made extensive contributions to the subject of number theory. Of these, none is more famous than that which occurred to him while leafing through the *Arithmetica*'s section on Pythagoras' last theorem.

Fermat wondered what would happen if instead of splitting a square into two squares, he tried to split a cube into two cubes. This amounted to looking for

How to become a celebrity mathematician

whole numbers a, b, c that satisfy the equation $a^3 + b^3 = c^3$. Geometrically, this would entail finding two cubes whose volumes add up to that of a third, with all the lengths given by whole numbers. Fermat could find no such numbers. Nor could he find any that satisfied $a^4 + b^4 = c^4$, or $a^5 + b^5 = c^5$, or indeed $a^n + b^n = c^n$ for any higher number n.

On the side of the page of his copy of *Arithmetica*, Fermat wrote that it was impossible to split a cube into two cubes, or a fourth power into two fourth powers, or any power higher into two like powers. Indeed, he infamously claimed to have a 'marvellous proof' of this fact, which the margin was unfortunately 'too narrow to contain'.

Wiles solves the unsolvable

Fermat was a truly great mathematician, but he did have a habit of claiming to have proofs where none are evident. Experts today do not believe that he could have had a complete proof of this fact. He did, however, manage to prove it in the specific case of fourth powers.

The general problem was to show that the equation $a^n + b^n = c^n$ could have no whole number solutions, for any $n \geq 3$. From when Fermat posed it in 1637, the problem survived for over 300 years. Despite the attentions of many of the greatest mathematical minds including Leonhard Euler, no-one could prove that no solution could exist. Was it possible that Fermat was wrong, and that an anti-Fermat triple would be found, contradicting his so-called *last theorem*?

In fact, Fermat was right, as many people had suspected. The matter was eventually laid to rest in 1995, by Andrew Wiles at the University of Cambridge. The equation would never be solved, as there are no whole numbers that satisfy it. Wiles' proof was a culmination of a monumental effort, in which the very landscape of mathematics shifted. Ultimately the proof resulted from building a bridge from the ancient subject of Diophantine equations to the modern world of *complex analysis* (see *How to admire a mathematical masterpiece*).

Fermat's last theorem for cubes

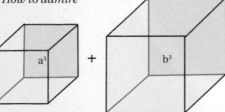
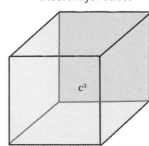

3 How to square a circle

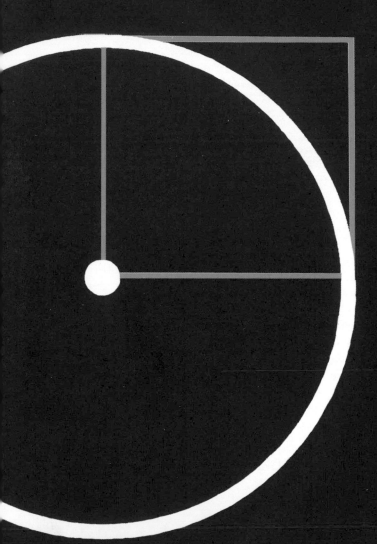

Perhaps humanity's first great technological triumph was the invention of the wheel. The enduring usefulness of this simple tool is down to the high level of symmetry of its shape: the circle. This has often been hailed as the ultimate in geometrical perfection. Certainly no other shape has captivated the imagination of technicians and geometers for longer.

What exactly is a circle? Around 300 BC, the father of geometry, Euclid of Alexandria, defined it this way: mark a point on a page. This will be the centre. Next, pick a distance, say 1 cm. Then mark every point on the page that is exactly 1 cm away from the centre.

The shape that emerges is a circle, specifically one whose radius is 1 cm. The *radius* of a circle is the distance from the centre to the curve. The *circumference* of the circle is the length all the way around it. A *diameter* of the circle is twice the radius: the distance from one side to the other through the centre. The relationship between these two quantities, the diameter and circumference, is the source of the longest running drama in mathematics.

● πr^2 and all that

An amazing number of questions are posed by this simplest of shapes. The first is: how long is it? That is to say, if we know its diameter, what is its circumference? On first impressions, it is not at all obvious what the answer should be. Starting with a circle of diameter 2 cm, if you measure the circumference you get approximately 6.28 cm. If the circle has a diameter of 4 cm, the circumference is around 12.57 cm. A diameter of 1 cm leads to a circumference of around 3.14 cm.

Looking at these numbers, ancient geometers realized that there seemed to be a fixed number that you could multiply by the diameter, to get the circumference. This number is known as π, the Greek letter pi. (The choice of name comes from the first letter of the Greek word *periphery*.) So if the circle has diameter d then its circumference is $\pi \times d$. In terms of the radius r, the circumference is $2 \times \pi \times r$, or simply $2\pi r$. For this formula to be useful, we have to know what the value of π actually is.

This question long predates the Greeks, and is among the most ancient of all mathematical topics. Over a thousand years earlier, Babylonian mathematicians had believed that the value of π was $\frac{25}{8}$, which is 3.125.

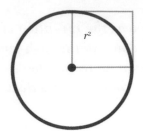

In Egypt the Rhind papyrus dating from 1650 BC uses an approximation of $\frac{256}{81}$ which is around 3.160. Both of these answers are close, but neither is exactly right. In fact, the full story would take millennia to be told. It would not be until the 18th century that mathematicians finally discovered that π is an *irrational* number, meaning that it can never be expressed exactly as a fraction.

Another important question one can ask about a circle is: what is its area? The answer is again provided by the number π. If the radius of the circle is 1, then the area of the circle is exactly π. More generally, if we build a little square on the radius of the circle, the number of times this fits into the whole circle is expressed by π. To express the same thing in a formula, if the circle has radius r, then the square built from the radius has area $r \times r = r^2$. This fits into the circle π times, meaning that the circle has area $\pi \times r^2$, or simply πr^2. So a circle with radius 2 cm has area $\pi \times 2 \times 2$, which is around $12.6\,cm^2$.

● The benefits of being square

Long before Euclid or Pythagoras, the mathematicians of ancient Egypt had already begun to investigate the relationship of the circle to that other mainstay of the geometric world, the square. While the secrets of the circle are shrouded in the mystery of π, the square is as straightforward as could be hoped. It is defined by a single number, its height. A square with height 3 has area $3 \times 3 = 9$. More generally, if the height of the square is h, then its area is $h \times h$, or h^2. In fact h^2 is pronounced 'h squared' in tribute to the simple arithmetic of this shape. At least, no mysterious irrational numbers seem to be required here. Perhaps this shape's predictability explains why 'square' became slang for 'boring' during the 1960s, as in 'be there – or be square!'

The question that has intrigued geometers since the time of the ancient Egyptians is that of *squaring the circle*. The question is this: begin with a circle. The challenge is to create a square that has exactly the same area.

We may as well suppose that the circle has radius 1 unit. That means that its area is $\pi \times 1^2 = \pi$. So what is needed is a square with area π. So the length of the side must be a number that when multiplied by itself comes to π. This is the square root of π, written √π.

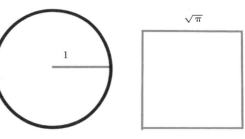

As Euclid showed, the square root does not present too much of a problem, it is the number π that poses the central difficulty. The question that so fascinated the ancient Greeks was whether it is possible to fashion this square using just the most elementary geometrical toolkit: a straight-edge and a pair of compasses.

Many interesting geometric structures can be built using only these simple tools. Euclid showed how to divide a line into two equal halves using these tools, a technique that is still taught in classrooms around the world. He also developed more sophisticated techniques, such as finding the square root of a number, and building regular pentagons using just a straight edge and compass. Ultimately, the question of squaring the circle reduces to this: given a line of length 1, can you construct a line of length π, using just a straight-edge and compass? This simple problem captured the very essence of π.

$$\underline{1} \quad \longrightarrow \quad \underline{\pi}$$

In defence of the irrational

It was one of Pythagoras' followers, Hippasus, who discovered that some numbers can never be written exactly as fractions. In fact, one of these strange *irrational numbers* even features in the dull, predictable square. If we have a 1 × 1 square, then its diagonal is √2, and this is an irrational number (see *How to become a celebrity mathematician*). There is no way to express it exactly as a fraction. This was a source of outrage to Pythagoras and, according to legend, Hippasus was drowned for this terrible heresy against the integrity of the whole numbers. Hippasus was right, though: irrational numbers are everywhere; they are unavoidable.

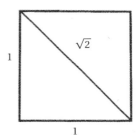

For many years it was suspected that π should be among the irrational numbers. But it was not until the work of Johann Lambert in the 18th century that this was finally established once and for all. A consequence of the fact that π is irrational is that if you write down its decimal expansion, which begins 3.14159265358979323846264. . ., it will never end, and nor will it ever repeat itself, it simply runs on for ever. (For this reason it makes an excellent basis for a memory test. There are many people who compete to memorize as many digits of π as they can; the world record officially stands at 67,890 digits, set by Lu Chao in 2005, although an unverified attempt by Akira Haraguchi in 2006 reportedly reached 100,000.)

Is π normal?

Because π is irrational we know that the decimal representation of π will never end, or repeat itself. That need not mean that there are no patterns in it. We might ask, for example, if each of the digits 0 to 9 occurs equally often or if some are more frequent than others. The question here is whether or not π is a *normal* number. Normal here carries a technical meaning: in a normal number, each of the digits 0 to 9 occurs with equal frequency, namely 10% of the time. More than this, however, all the possible pairs of digits from 00 to 99 must also occur with equal frequency (1% of the time). Similar requirements hold for triplets of digits, quadruples and longer formations. This means that there is no overall trend or tendency within the string of digits.

Actually, the requirement of normality is even more stringent than this. The ordinary representation of π as 3.141592653. . . is based on the *decimal* system, that is, our usual way of writing numbers that is based on the number 10. However, any other whole number can equally well function as a base for writing numbers (see *How to talk to a computer*). In binary for example, π is written as 11.001001000011111101101. . .

To be normal the same unpredictability must apply in every possible base. It has long been suspected that π should be normal, but even today this remains unproved.

Transcendence: the final conquest of π

The irrationality of π was a breakthrough, but it was not enough to settle the question of squaring the circle. The problem is that some irrational lengths can be constructed using ruler and compass. One such is $\sqrt{2}$. After all, it is easy to draw a 1 × 1 square using a ruler and compass, and, as Hippasus realized, its diagonal has length $\sqrt{2}$. This shows that the numbers that can be constructed are not limited to the rational numbers; some irrational numbers are constructible too. Perhaps π was one of them?

The answer was no. For π is an even more exotic type of number than merely irrational. It is *transcendental*, which means that, however much you add π to itself, or multiply it by itself, you will never return to the comfortable home ground of the whole numbers. This is in contrast to $\sqrt{2}$ which, when you multiply it by itself, does produce a whole number (2). Transcendental numbers such as π never do. So $\pi \times \pi$ cannot be a whole number, and nor can $5\pi^3 + 4\pi^2$, or $3\pi^{17} + \pi$, or any of the endless variations on this theme. No matter how much you multiply π by itself, and add the results together, you can never build a bridge to the safety of the whole numbers. Arithmetically speaking, π will forever remain out of reach.

Transcendental numbers were first discovered in the 19th century, by Joseph Liouville. He wrote down a number 0.110001000000000000000001000. . . that had 1s at the 1st, 2nd, 6th, 24th, . . . positions and 0s everywhere else. (The pattern for the positions of the 1s is 1, 2 × 1, 3 × 2 × 1, 4 × 3 × 2 × 1, and so on.) Mathematicians were intrigued by Liouville's strange number, but did not expect that the phenomenon he had identified would have such an important role in mathematics. It was Ferdinand von Lindemann who, near the end of the 19th century, proved that π is also transcendental.

Earlier, Pierre Wantzel had proved that no transcendental number could ever be constructed by ruler and compass. Together these two breakthroughs finally settled the ancient question of squaring the circle: it can never be done. Right at the heart of the simplest of all geometrical shapes, lies an unconstructible, transcendental number, forever out of our reach.

4 How to win the ultimate mathematics prize

- Prime numbers
- Bertrand's postulate and twin primes
- The prime number theorem
- Bernhard Riemann's hypothesis

Mathematicians study all manner of exotic objects, from different denominations of infinity to million-dimensional hypercubes, but, above all else, they study numbers. As such, prime numbers are the most fundamental objects in the whole subject, as they are the numbers from which all others are built. The definition is not complicated: a prime number is defined to be a whole number that cannot be divided by any other. So 6 is not a prime number, as it can be divided by 2, but 7 is prime, as no other number divides it (except itself and 1, which we always allow). Primes have been studied for millennia but, even with the power of modern mathematics, and high-speed computer processors, they continue to guard their secrets closely.

What is so important about these numbers? The answer is that primes are the building blocks from which all other numbers are built. This fact goes by the grand title of the *fundamental theorem of arithmetic*. If we start with a whole number such as 12, it can be broken down into prime numbers: $2 \times 2 \times 3$. What is more, there is only one way of doing this; multiplying different primes together will always produce a different answer.

To understand numbers then we must understand this sequence of numbers:

$$2, 3, 5, 7, 11, 13, 17, 19, 23, 29, \ldots$$

The first major insight into prime numbers was set on paper by Euclid in his book *The Elements*. Although best known for its geometry (see *How to draw an impossible triangle*), *The Elements* also included crucial insights into number theory. The most important was that this list of primes goes on forever; it never ends. No matter how large a prime number we discover, there will always be a bigger one.

At the time of writing, the biggest prime we know is $2^{43,112,609} - 1$. (That is 2 multiplied by itself 43,112,609 times, minus 1. To write it out fully would require several volumes, since it is 12,978,189 digits long). It was discovered in 2008 as part of the Great Internet Mersenne Prime Hunt. Thanks to Euclid, though, we know that there remains a never-ending list of even larger primes, waiting to be discovered (or forever out of our reach).

The infinity of the primes has an important consequence. Mathematicians want to understand the patterns that exist within the prime numbers. If there were only finitely many, we could simply look at them and draw conclusions. But since there are infinitely many, we need a more abstract approach. Number theorists have developed many sophisticated techniques; but the primes are tough customers and many questions remain unanswered.

The gaps between the primes

How far apart can it be from one prime to the next? The primes start off coming thick and fast: 2, 3, 5, 7, . . ., but later the gaps grow longer: 199, 211, 223. As we progress along the list of primes, might we suddenly find a gap of several million? The answer is yes, but not for a while. The size of the gaps can never grow disproportionately bigger than the size of the primes themselves. This was an observation made in 1845 by Joseph Bertrand. He said that if you take any number (call it n) and double it ($2 \times n$, or $2n$ for short), you must always be able to find at least one prime number in between the two. So we can rest assured that there must be a prime number between 1 billion and 2 billion, for example. This can save a lot of effort, as pinning one down exactly is difficult. Bertrand was not able to provide a proof for his claim, but he was right. *Bertrand's postulate*, as it is called, was proved in 1850 by Pafnuty Chebyshev.

A more recent conjecture made by Dorin Andrica says that the true bound should be much tighter than Bertrand suggested. Andrica's conjecture holds that the maximum gap between primes is of the order of $\sqrt{n}$, rather than n. However, no proof has yet been found for this.

Pairs of primes

Some primes are very close to each other. The very first primes, 2 and 3, are one immediately after the other. Might there be others like this? The answer is no. The reason is that if you take any two successive numbers, then one of them must be even and the other odd. Even numbers cannot be prime, because they are divisible by 2 (the only exception is 2 itself). So prime numbers cannot in general sit next to each other.

The smallest recurring gap is between primes that sit 2 apart. Some examples are 3 and 5, 17 and 19, 59 and 61, 881 and 883. These are known as *twin primes*. There is a long list of them; the largest currently known pairs are many thousands of digits long. However, we do not yet know whether the list of twin primes is in fact infinite.

The *twin prime conjecture* holds that it is: there is no biggest pair of twin primes; you can always find a bigger one. This has yet to be proved, and is symptomatic of the secretive nature of prime numbers.

'Chebyshev said it, but I'll say it again; there's always a prime between n and 2n.'

NATHAN FINE

How to win the ultimate mathematics prize

Gauss zooms out

There are many conjectures, such as the twin primes conjecture and Andrica's conjecture, which consider the permissible sizes of gaps between primes. There are rather fewer concretely established theorems, but Bertrand's postulate is one of them. All of these questions focus on the *local* behaviour of individual primes. If we are at one prime, how far away must the next one be? A different approach is to zoom out, and view the distribution of the primes over a much broader range.

This is an extremely rewarding exercise. On a small scale, the glitches and jumps make the prime numbers very unpredictable. On a broader scale, however, clear patterns emerge. The most important pattern was noticed by Carl Friedrich Gauss in 1792, at the age of just 15. What the young Gauss considered was the *prime counting function*. Pick a number, such as 10 or 200. Then count the number of primes up to our chosen number. So, for example, there are four primes up to 10, namely 2, 3, 5, 7. Up to 200, there are 46.

The question Gauss addressed is: how fast does this counting function grow? Of course every time you reach a prime, the count jumps up by 1, so this is no more predictable than the individual primes themselves. However, what interested Gauss was how fast this grew on average, over a large range. Up to a million, around how many primes should we expect? Would it be nearer a hundred or a hundred thousand? Gauss was not aiming for a precise answer, but an estimate. As it happens, he was able to provide a very good one indeed.

The flight-path of the primes

The *prime number theorem* that Gauss had glimpsed would become one of the central results in the subject. Gauss' hunch was that the number of primes between 1 and n is approximately equal to $\frac{n}{\ln n}$, where $\ln n$ is the *natural logarithm* of n.

Natural logarithms had been understood for many years and are not complicated. They are based on the number e (see *How to admire a mathematical masterpiece*). Essentially, $\ln b = a$ means that $e^a = b$. No-one had expected that they should play such a central role in unravelling the mysteries of the prime numbers.

The natural logarithm of one million is around 13.8155 (to four decimal places), because $e^{13.8155} = 1,000,000$ (approximately). So, according to Gauss' estimate, the number of primes between 1 and a million should be approximately $\frac{1,000,000}{13.8155}$, which is 72,382 to the nearest whole number.

'I think prime numbers are like life. They are very logical but you could never work out the rules, even if you spend all your time thinking about them.'

MARK HADDON

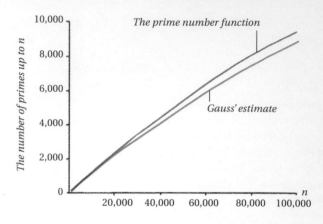

In fact the exact number of primes up to one million is 78,498 (starting at 2 and ending at 999,983). So the result is not perfect, but as an order-of-magnitude estimate, it is not bad. Gauss believed that, as we look at ever bigger numbers, his estimate should get proportionally closer and closer to the true value. In fact, he went on to provide an even better estimate, by slightly altering the natural logarithm. However, even in adulthood and at the peak of his powers, he was not able to prove that his estimates were accurate. That would have to wait for another major leap in the theory of prime numbers.

Riemann's perfect map

Gauss' prime number estimate was tantalizing. But, to be of value, it first needed to be proved. After proof, an even better prospect would be to improve the theorem, to provide a formula giving the *exact* number of primes up to a certain limit, a perfect map of the prime numbers. This seemed far too much to hope for. But, even before Gauss' estimates had been proved, Bernhard Riemann's 1859 paper 'On the number of prime numbers less than a given quantity' provided exactly such a map. It was a stunning achievement, a landmark in the history of number theory, and a technical tour de force. There was only one problem. Riemann's theorem relied on detailed information about a highly mysterious object at the centre of his formula, now called the *Riemann zeta function*.

A *function* is like a machine that takes numbers as inputs and gives out other numbers as outputs. A simple function *f* might square every input *s*, for example. So the defining formula of that function would be $f(s) = s^2$, and particular examples would be $f(3) = 9$ and $f(5) = 25$. Riemann's function is more complicated than this. Technically, it is defined by taking a number *s* as the input, and giving an output of $\zeta(s)$, defined as:

$$\zeta(s) = 1 + \frac{1}{2^s} + \frac{1}{3^s} + \frac{1}{4^s} + \frac{1}{5^s} + \cdots$$

This definition is already very intricate, but it is not quite adequate for all numbers. Riemann had to extend this function, to allow every possible *complex number* to be included as an input (see *How to solve every equation there has ever been*). Through some ground-breaking work, Riemann was able to find a way to do this. The critical question was where this extended

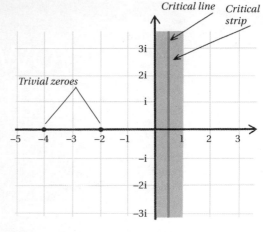

Trivial zeroes

function vanishes; he needed to know which input values produce an output of 0, that is, for which values of s is $\zeta(s) = 0$. These values are known as *Riemann's zeroes* and are the key to our understanding of the prime number function.

There are some values where it was comparatively easy to see that the function was zero: -2, -4, -6, -8, and so on. These are known as the *trivial zeroes* of the Riemann function. The crux of the problem revolves around the others, the non-trivial zeroes. Riemann was able to see where they should lie. He thought that they should all sit on a particular vertical line in the space representing complex numbers. Known as the *critical line*, it is defined as those numbers whose real part is $\frac{1}{2}$. So all of the zeroes should have the form $\frac{1}{2} + ix$ for different values of x. This is the famous *Riemann hypothesis*.

Riemann was not able to prove his hypothesis, but wrote that he thought it 'very probable'. However, even without the hypothesis, his zeta function was a powerful new tool for studying prime numbers. In 1896, Jacques Hadamard and Charles de la Vallée Poussin managed to narrow down the region where the zeroes might exist to within a certain *critical strip*, and in so doing were able finally to prove Gauss' prime number theorem.

The toughest question in mathematics

Bernhard Riemann's hypothesis has defied 150 years of attempted proofs. As it holds the key to our understanding of the prime numbers, it surely remains the greatest open question in mathematics today. In 1900, the mathematician David Hilbert formulated a list of 23 problems that set the direction of mathematics in the 20th century; the Riemann hypothesis was among them. By the dawn of the 21st century, it had still not been proven. In 2000, the Clay Mathematics Institute put together a list of seven Millennium problems, each carrying a $1,000,000 prize. One of these was the Riemann hypothesis. The money remains unclaimed today.

'If I were to awaken after having slept for a thousand years, my first question would be: Has the Riemann hypothesis been proven?'

DAVID HILBERT

5 How to slay a mathematical monster

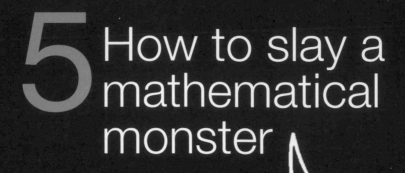

- The science of symmetry
- Group theory takes over the world
- The classification of finite simple groups
- The monster and its moonshine

Why are some faces more beautiful than others? Part of the answer is that humans are instinctively attracted to symmetry; it is in our DNA. Symmetry runs throughout our art and civilization, from the robust fourfold symmetry of the Egyptian Pyramids to the symmetric patterns on wallpapers and carpets today (see How to design the perfect pattern).

Symmetry is also a huge topic of research in mathematics, where it goes by the name of *group theory*. The central objects here are abstract entities called *groups*, which beautifully capture the essence of symmetry. The total number of possible groups is vast and unmanageable, and the attempt to navigate this ocean of symmetry has produced some of the most astounding mathematical theorems yet proved. Perhaps the most spectacular is the *classification of finite simple groups*. As well as providing a navigable chart to the ocean of symmetry, the classification revealed an extraordinary creature that lives there, known to mathematicians simply as *the monster*.

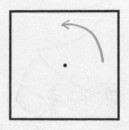

The science of symmetry

Start with a square. If we rotate it by 90° anticlockwise, it looks the same as when we started. So rotation by 90° is called a *symmetry* of this shape. This is a slight difference from ordinary language, in that a symmetry is thought of as an *action*, something you can do to a shape that leaves it looking the same. What are other symmetries of the square? We could rotate it by 180° or 270°. *Rotational symmetry* is exploited in the natural world. Bees and other insects that plants need for pollination are attracted by the rotational symmetry of the plant's flowers.

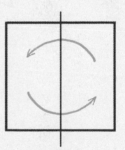

There is another way in which a square is symmetric. If we draw a vertical line through the centre of the square, and then reflect it across the line, again it ends up looking the same. A horizontal line also works as a mirror, as do the two diagonals. This gives a total of four reflections, on top of the three rotational symmetries. In fact, the square has one more symmetry: we can just leave it alone. Even the most irregular shapes have at least this one trivial symmetry. Taken together, these eight symmetries make up what is called the square's *symmetry group*. More complex shapes may have bigger groups. In three dimensions, the cube has 24 symmetries, for example.

Two symmetries of a square

The laws of symmetry

Just counting the symmetries of a shape does not tell us much about it. The square has eight symmetries, but so does the eight-pointed curvy star illustrated overleaf (the curves mean that there can be no reflectional symmetry, only eight rotations). As well as counting the symmetries, there is a subtler observation we can make. This is that symmetries can be combined: we can perform one action after the other. So what happens when we rotate the square by 90°, and then reflect it in the vertical line? To answer this question, we label the corners of the square, temporarily. The answer is that the result is the same as reflecting the square once, in one of the diagonal lines (see opposite).

This is not obvious, and goes to the heart of what a mathematical *group* is. There are three key observations to make:

1. Whenever you combine two symmetries, you always get a third.

2. There is one special symmetry, the identity, which has no effect at all.

3. Every symmetry has an opposite or *inverse*, which cancels it out. For example, the inverse of rotating by 90° is rotating by 270°.

These rules are reminiscent of something else, from a seemingly unrelated area of mathematics, namely adding whole numbers.

1. Whenever you add two whole numbers, you always get a third.

2. There is one special number, 0, which has no effect at all.

3. Every number has an opposite or *inverse*, which cancels it out. For example, the inverse of 5 is −5, because these combine to give 0.

Mathematicians get excited when they see similar rules at work in different contexts. Their instincts are always that the point of similarity must be important. In this case, the symmetries of the square and the addition of numbers are both examples of a group, essentially defined as any collection of objects, together with some rule for combining them, which satisfies these three laws. (There is an additional requirement for the process of combination, specifically that when you combine three objects it should not matter whether you combine the third with the combination of the first and second, or combine the first with the combination of the second and third. For the whole numbers an example is $(2 + 3) + 4 = 2 + (3 + 4)$, which is why we can write $2 + 3 + 4$ without having to worry about any ambiguity.)

Groups come in two principal flavours. The symmetry group of the square is *finite*, as it contains just eight different possible actions. However, there are infinitely many whole numbers, making this an example of an *infinite group*. Both finite groups and infinite groups are widespread in the world. The symmetries of a circle form an infinite group because you can rotate it through any possible angle and leave it looking the same.

Group theory takes over the world

The study of groups got under way in the early 19th century, with the pioneering work of Niels Abel and Évariste Galois. These mathematicians found the same group laws at work in yet another realm of mathematics.

One of the oldest tasks that mathematicians have sought to accomplish is solving equations (see *How to solve every equation there has ever been*). Essentially this is a guess-the-number game: I think of a number x, and multiply it by itself and then subtract 4. If I am left with 0, what was the number I was first thinking of? This can be expressed as an equation: $x^2 - 4 = 0$. To *solve* this equation is to find all the possible values of x for which this assertion is true.

What Abel and Galois realized is that, just as a square has a group of symmetries that swap around its corners but leave the overall shape looking the same, every equation has an underlying group that swaps around its solutions, but leaves the equation looking the same. This curious insight was turned into a wonderful theorem, to do with the solvability of equations.

How do you solve an equation? For some equations there is a simple way to proceed. A formula drilled into the heads of generations of maths students is:

$$x = \frac{-b \pm \sqrt{b^2 - 4ac}}{2a}$$

This provides a method for solving any quadratic equation, meaning one like $x^2 - 4 = 0$, which contains x^2 but not x^3 or higher terms. All that is needed is to substitute in the values of a, b and c, according to $ax^2 + bx + c = 0$. What about more complicated equations, which involve x^3, x^4, and higher terms? During the 16th century, Italian mathematicians found more complicated formulae that could also solve these.

Abel and Galois realized that the solvability of an equation depends in a very precise way on the structure of the underlying group. In particular for quintic equations, that is those involving x^5, there can be no simple formula for solving them, as the underlying group is irreducibly complex.

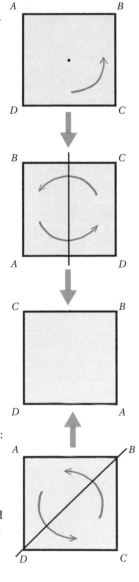

Symmetries at work

Simple groups: the atoms of symmetry

The insights of Abel and Galois showed that the new abstract approach to groups could pay dividends. Later in the 19th century, the study of groups in general, and finite groups in particular, developed into a major industry. A critically important fact was discovered by Camille Jordan and Otto Hölder.

A prime number is defined to be one that cannot be broken down as a product of smaller numbers. For example, 7 is prime, while 8 is not prime, because 8 = 2 × 4. Similarly, some groups can be broken down into smaller ones, while others cannot. These indecomposable groups are known as *simple* (extremely misleadingly, as later became clear). The Jordan–Hölder theorem established the importance of these simple groups. It showed that, just as any number can be broken down into a product of prime numbers (for example 6 = 2 × 3), so any finite group can be broken down into simple groups. What is more, this can be done in only one way, just as a number can be broken down into prime factors in only one way.

The implication of Jordan and Hölder's work was clear: to understand finite groups, the first step was to understand the possible simple groups, the atoms from which the others were built. This turned into a truly monumental task, which would finally be accomplished in one of the milestones of 20th-century mathematics.

The classification of finite simple groups

Some mathematical theorems emerge unexpectedly, the result of the insight and hard work of one person working single-handedly. The classification of finite simple groups was not like this. Instead it was approached incrementally by mathematicians around the world, all trying to find some pattern in the bewildering array of finite simple groups that were continually being discovered. This team effort was coordinated by Daniel Gorenstein, who in 1972 had the vision to see that proof might finally be attainable.

By 1981, through the concentrated efforts of hundreds of people, a picture had gradually emerged. It turned out that there are exactly 18 different families of finite simple groups. These are the symmetry groups of abstract geometrical structures, similar to that of the square. Each family contains infinitely many groups (just as we can have a symmetry group of a shape with 4, 5, 6 or 100 sides, or as many as we like). What the classification theorem aimed to do was show that every conceivable finite simple group must be of one of these 18 types. However, the story turned out to have another twist.

The entrance of the monster

As well as pinning down the 18 families, group theorists had found several awkward individual groups that did not fit into any of them. This made the classification project extremely difficult, as there seemed no reason why there could not be a never-ending stream of ever larger, more complicated simple groups that refused to fit any pattern.

Happily, this turned out not to be the case. It was eventually determined that there are exactly 26 of these *sporadic groups*. The largest of them is a formidable creature known as *the monster*. While the symmetry group of the square contains just eight elements, the monster weighs in at a hefty 808,017,424,794,512,875,886,459,904,961,710,757,005,754,368,000,000,000.

The existence of the monster was predicted in 1973 by Bernd Fischer and Robert Griess, and was constructed by Griess in 1980. Taming this creature was a major challenge to the classification: could they be sure that there was just one, rather than a family of monsters? The question was answered, and the project completed, in 2004. The final theorem says that any finite simple group must either be a member of one of the 18 families, or one of the 26 sporadic groups.

Monstrous moonshine

The classification of finite simple groups is one of the triumphs of mathematics in the modern age, and it has an equally wonderful postscript. In a seemingly unrelated area of mathematics, in the deepest waters of modern complex analysis (see *How to admire a mathematical masterpiece*) is the realm of *modular forms*. These powerful but secretive objects have played a prominent role in mathematics in recent years, including in the proof of Fermat's last theorem (see *How to become a celebrity mathematician*). They seem to have nothing whatsoever to do with finite groups, and yet, in 1979, John Conway and Simon Norton spotted the tell-tale claw-marks of the monster here. Numbers such as 196,883 and 21,493,760, which described aspects of the monster, also appeared in this alien world. They dubbed this phenomenon *moonshine*. The connection was finally made explicit by Richard Borcherds in 1992, in work for which he won the Fields medal.

6 How to excel at Sudoku

- As easy as 1, 2, 3
- Graeco-Latin squares and the 36-officers problem
- Sudoku sweeps the world
- The art of correcting errors

1	2	3
2	3	1

The puzzles known as Latin squares have an illustrious mathematical history. Known as wafq majazi in Arabic, the first examples come not from ancient Rome as their name suggests, but from engravings on good luck charms in the 13th-century Islamic world. Within pure mathematics these same squares appear as Caley tables, the times-tables of abstract forms of multiplication known as groups (see How to slay a mathematical monster). In applied mathematics, they are extremely useful as error-correcting codes. Even more famously, however, Latin squares feature daily in newspapers around the world, under the name of Sudoku.

As easy as 1, 2, 3

To create a Latin square, draw a 3 × 3 table. The task is then to write in the numbers 1 to 3 in such a way that each digit occurs exactly once in each row and in each column. This is easy to do, and there are several possible answers. Here is one:

1	2	3
2	3	1
3	1	2

Of course, this challenge then extends to 4 × 4, 5 × 5 squares, or as large as you like. This is the basic idea of Latin squares, and, since they were discovered by medieval scholars, generations of people have enjoyed the economy and symmetry of their design. Adding extra requirements to it produces new phenomena such as Sudoku, and Graeco–Latin squares.

The difficulties of staging a military tattoo

When the Swiss mathematician Leonhard Euler first heard about Latin squares, they set his great mind racing. He immediately looked at ways to enrich them with extra conditions. The first thing he came up with was a way to fit two Latin squares together.

Suppose that nine military officers are going to march in square formation, as part of a Royal Parade. There are three different regiments represented (call them A, B and C), with three officers from each. To maximize the symmetry, the general instructs the officers to arrange themselves so that, in each row and column of the square, each regiment is represented exactly once. This amounts to the officers forming a Latin square. The extra condition comes from the fact that there are also three different ranks represented (call them 1, 2 and 3, the same in each regiment). The general also wants each rank to be represented exactly once in each row and column. This means that the officers have to arrange themselves in a Latin square, in two ways simultaneously: first from the perspective of regiment, and secondly from the point of view of rank.

Then there are nine officers in total represented as A1, A2, A3, B1, B2, B3, C1, C2, C3. Now we can solve the problem:

A1	B2	C3
B3	C1	A2
C2	A3	B1

This is an example of a Graeco–Latin square. It comes from gluing together two different Latin squares:

A	B	C
B	C	A
C	A	B

and

1	2	3
3	1	2
2	3	1

Any two Latin squares will not do, however. The result must be such that no single pairing (such as A2) is repeated, and all are represented. (The term Graeco–Latin square reflects the fact that Euler used the Latin and Greek alphabets as the two sets of symbols.)

The nine-officers problem amounts to finding a 3×3 Graeco–Latin square, and is fairly simple to accomplish. If you try the same thing with a 2×2 square, it is easy to see that it is impossible, meaning that there is no solution to the four-officers problem. Larger 4×4 and 5×5 Graeco–Latin squares are trickier to construct, but can be done with a little trial and error, thereby solving the problems of 16 and 25 officers. Altogether more troublesome is the famous 36-officers problem. The question is the same: is it possible to find a 6×6 Graeco–Latin square? However, in this case, an answer is much more difficult to find. Euler finally wrote that 'after spending much effort to resolve this problem, we must acknowledge that such an arrangement is absolutely impossible, though we cannot give a rigorous proof'.

Euler was also unable to find Graeco–Latin squares of size 10×10 and 14×14, although the intermediate sizes seemed unproblematic. He therefore conjectured that no such squares can exist for any of the sizes 6, 10, 14, 18, 22, 26, and so on. (The pattern is that these are the numbers that are 2 more than a multiple of 4.)

The impossible officers

Was Euler right that the 36-officers problem had no solution? Settling this conjecture was a difficult process. The only possible approach seemed to be to list all possible 6 × 6 Latin squares, and see whether any two of them could be fitted together in the right way. This indeed was how the problem was eventually settled. In 1901, Gaston Tarry produced an exhaustive compendium of all the 9,408 possible 6 × 6 Latin squares, and established that no two of them could be combined to form a Graeco–Latin square.

Euler was right about the 36-officers problem: there is no possible arrangement that can satisfy the general. However, he was wrong about the larger squares. 10 × 10 Graeco–Latin squares can exist, as can 14 × 14 and all larger-sized squares. This was shown in 1959, by Parker, Bose and Shrikhande. So it is only squares of size 6 × 6 that are impossible.

Designing experiments and sports contests

An enduring use for Latin squares is in designing scientific experiments to minimize bias. Suppose we have five cars (A, B, C, D, E) to test for speed and safety, and five drivers to test them (call them 1, 2, 3, 4, 5). We could just assign each driver one car and compare results at the end, but that would very likely lead to bias, as the different drivers could well have different styles and skills. So we really want each driver to test all the cars, after which we will take the average time for each car and compare these. A suitable arrangement for this is given by a 5 × 5 Graeco–Latin square:

'The one thing I never want to see again is a military parade.'

GENERAL ULYSSES S. GRANT

	Track 1	Track 2	Track 3	Track 4	Track 5
Session 1	A1	B2	C3	D4	E5
Session 2	E2	A3	B4	C5	D1
Session 3	D3	E4	A5	B1	C2
Session 4	C4	D5	E1	A2	B3
Session 5	B5	C1	D2	E3	A4

This has many advantages. For example, there might be a disadvantage to going first, but each car will be the first for one driver, so this is spread out. The same approach is useful for scheduling, both in computer science, and for events such as sports contests. Suppose two teams of five sprinters are going to compete against each other. The square above provides a possible schedule for all the necessary races.

Sudoku sweeps the world

In New York in 1979, Howard Garns set off a chain of events that would change the world forever, at least in the eyes of people who enjoy brain-bending puzzles. He invented a new type of number game, which was published by Dell Pencil Puzzles and Word Games magazine under the name of *Number Place*. It was later taken up in Japan, where it became popular under the name '*Suuji wa dokushin ni kagiru*', meaning 'numbers should be unmarried', or *Sudoku* for short.

The basis of Sudoku is a 9×9 Latin square. The digits 1 to 9 have to be inserted into the grid in such a way that each one appears exactly once in each row and in each column. However, there is one extra rule. The 9×9 grid is subdivided into nine 3×3 squares. Each such square must also contain the numbers 1 to 9. You can start from an empty grid, but typically the puzzle comes with a few numbers already filled in as clues. The challenge is then to complete the rest of the grid.

The clues to the Sudoku have to be chosen very carefully, a surprisingly tricky mathematical procedure. Firstly, it must be done in such a way that the puzzle has a solution. It must be possible to complete the grid, satisfying all the rules. This is not straightforward, as contradictory configurations are often not immediately obvious. More subtly, there should be only one solution. Starting with the clues, it should be possible to proceed through logical deduction alone, to arrive at the only possible answer. This is what mathematicians call an *existence* and *uniqueness* problem.

An interesting question regarding Sudoku is: what is the minimum number of clues needed to guarantee that there is exactly one solution? Surprisingly, despite the worldwide attention these puzzles have received, the answer is still unknown, but widely suspected to be 17. Certainly it is not any bigger than this. If you could prove that 17 is the minimum possible such number, or alternatively find a configuration of 16 clues that guarantee a unique solution, you would truly have earned Sudoku grandmaster status (even if you can't solve the pesky things).

The art of correcting errors

Here is a 5 × 5 Latin square with a mistake in it. Can you spot the error?
(The answer is below.)

2	4	1	3	5
3	5	2	4	1
4	1	3	5	2
1	3	5	2	5
5	2	4	1	3

The fourth row as two 5s in it, and no 4s. So, one of the 5s should be a 4.
But which one is it? We can compare the two relevant columns. The third
column is fine, but the final column also has two 5s in it, and no 4s. So it
must be that the 5 in the fourth row, final column should be a 4.

This goes to show that Latin squares have a remarkable property: even
when an error creeps in, it can not only be identified, but actually corrected,
very easily. The line of reasoning above can easily be automated and
performed by a computer. This is extremely useful, as errors are an
ineluctable part of life.

In the information age, the ability to identify and correct errors
automatically is highly prized. For example, when radio is broadcast
digitally, some errors are bound to creep in due to interference with other
signals and ambient noise. By encoding Latin squares into the data-stream,
it is possible to eliminate these errors after the signal has been received.

Error correction is the heart of Sudoku, and it shows the high level of
accuracy that is possible. The point of the puzzle is that the entire grid of
81 numbers can be perfectly reconstructed from just a small proportion of
them, 17 or so.

By building Sudoku and similar structures into streams of binary data, even
if a large proportion of the signal is lost or corrupted, it can still be possible
to reconstruct the entire message with near-perfect accuracy (see *How to
talk to a computer*). Indeed, in 1948 Claude Shannon proved the fundamental
result that there is no limit to how much information can be lost, and still
allow the message to be reconstructed with great accuracy. Remarkably, he
showed that there are ways to encode the data so that this can be done to any
desired level accuracy, without making the message unworkably long.

'*Be not
ashamed of
mistakes and
thus make
them crimes.*

*A man who
has committed
a mistake and
does not
correct it, is
committing
another
mistake.*'

CONFUCIUS

7 How to unleash chaos

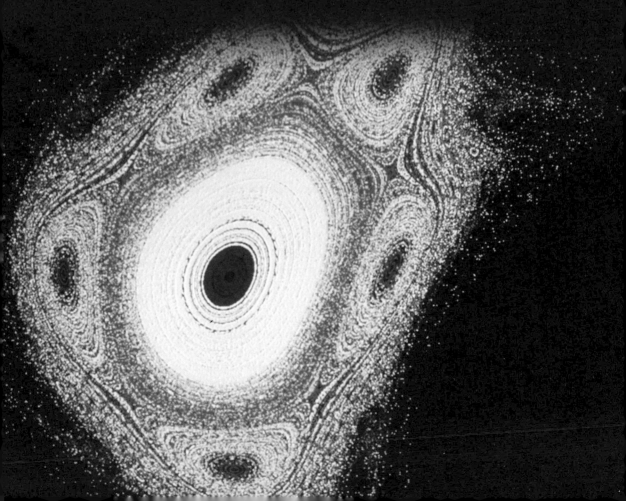

How is science possible? The answer is that we live in a world that is ordered. How things have behaved in the past is a good guide to how they will behave in the future. The sun rises each day; when I drop a ball, it falls to the ground. This allows us to make and test predictions: scientists can tell the future. The enemy of order is chaos. In a totally chaotic universe, science would not be possible, as the current state of things would be useless as a guide to the future.

The universe we inhabit hovers somewhere between the two poles of chaos and order. There are chaotic processes at work around us, a great many of them. At the same time, they sit within a broader framework of order. It is this delicate balance that makes our part of the universe fertile ground for scientific enquiry, and indeed for life itself.

The infinite peg-board

If I drop a ball to the ground, I can apply the insights of past scientists to get a good idea of where and when it will land. Of course this prediction will not be perfectly accurate. I cannot expect to identify the exact landing spot, down to the nearest nanometre. Why not? Well, the primary reason is that I do not know exactly where it started, or at exactly what time it was dropped. So there is a little uncertainty in the outcome, but it is within acceptable bounds. Most importantly, the level of uncertainty about the eventual outcome matches that uncertainty about the starting conditions; nothing happens in between to increase it.

If instead I drop the ball onto a knife-edge, something new occurs. Which side the ball lands depends much more closely on the exact position of release. A small variation one way or the other can make a significant difference to the outcome. At least there are only two possibilities to worry about, however. If I cannot tell exactly which side of the knife the release-point is, I can at least estimate the probability of each of the two possible outcomes, at 50% each. So there has been some increase in the uncertainty of the outcome.

The situation is worse when we drop the ball onto a peg-board. At each level, the ball may fall on either side of the peg. Its route and its eventual resting place now depend on the starting conditions very closely indeed. What mathematicians call *chaos* resembles an infinitely long peg-board. If there is any uncertainty whatsoever about the ball's initial position (and in any practical application there always must be), then its ultimate route is unknowable. Any change, no matter how tiny, will fundamentally change its course. The slightest uncertainty about the initial position of the ball explodes it into total ignorance of the eventual outcome.

● A × B × C = chaos

It may sound like an oxymoron, but some chaotic systems are actually very simple. The simplest example of a chaotic mathematical process is called the *logistic map*. All that is involved is multiplying three numbers together. We call the *input x*, which will be somewhere between 0 and 1. This is analogous to the ball's starting position. We might start with an input of 0.25. The second number is $1 - x$, which in this case will be 0.75. The third number is the key to the whole system. It is called the *parameter* and is written as *r*. This value stays fixed for the duration of the process, and describes the layout of the peg-board, as it were. We will start by considering the system that has $r = 2$. Then the input is *x*, and the output is $r \times x \times (1 - x)$. We might write this as:

$$x \mapsto r \times x \times (1 - x)$$

When we start the process with an input of 0.25, we multiply the three ingredients together and get $2 \times 0.25 \times 0.75 = 0.375$. This value of 0.375 is the first output.

To continue the process, we feed this output back in as an input. This again means combining the three ingredients: *r* (that is 2), the new input 0.375, and $1 - 0.375 = 0.625$. This gives the next output as $2 \times 0.375 \times 0.625 = 0.469$ (to 3 decimal places).

Now, if we put this back in as an input, we will get an output of 0.498 (to 3dp) followed by a number very close to 0.5. So running this process has produced a sequence: 0.25, 0.375, 0.469, 0.498, 0.500, . . . As the process runs, the sequence gets ever closer to the number 0.5, called an *attracting point* of the system. In fact, the original input of 0.25 has little impact on the eventual outcome. No matter where we start the system (as long as it is between 0 and 1), the sequence will always home in on its attracting point: 0.5. This is the mathematical equivalent of dropping a ball on to flat ground. When we change the value of *r*, however, the situation changes in dramatic fashion.

● The forked road to chaos

The interesting question is what happens to the logistic map when we alter the value of *r*. If we change *r* to 3.3, and start the process again with an initial

input of 0.25, rounding off the numbers to two decimal places, we get a
new sequence:

0.25, 0.62, 0.78, 0.57, 0.81, 0.51, 0.82, 0.48, 0.82, 0.48, 0.82, 0.48, 0.82, . . .

As the process runs, this time it does not settle down to a single fixed point,
but rather fluctuates between two numbers near 0.48 and 0.82. This is called
an *attracting 2-cycle* of the system.

If we increase the value of r again to 3.5, and restart the process we get a
new sequence, and this time it has an *attracting 4-cycle* around 0.50, 0.87,
0.38, 0.83.

In fact, as we consider new systems with the value of r increased, we find
attracting 8-cycles, 16-cycles, 32-cycles, and so on. This is known as a
sequence of *bifurcations*.

After a certain point this forking ceases. Once the value of r exceeds 3.57, we
have a completely new phenomenon. Suppose we take $r = 3.7$ and restart the
process, we get:

0.25, 0.69, 0.79, 0.62, 0.87, 0.42, 0.90, 0.33, 0.82, 0.55, 0.92, . . .

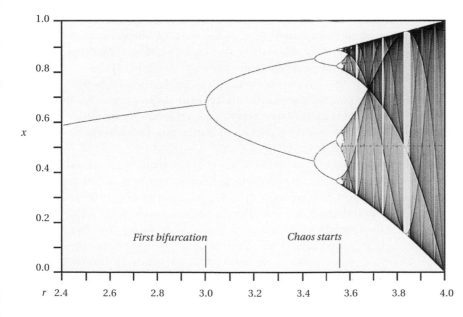

*The bifurcations
of the logistic
map*

It is tempting to believe that if we let this sequence run for long enough, some pattern will eventually emerge. In fact, it will not. This is the essence of mathematical *chaos*. The sequence simply jumps around, seemingly at random, however long we run it for. What is more, if we change the initial input even by a microscopic amount, the new sequence will end up looking totally different from the original.

The butterfly effect

The chaos of the logistic map is a fascinating phenomenon for mathematicians to puzzle over, but is it any more than that? The defining trait of chaos is its *sensitivity to initial conditions*. Even the tiniest alteration to the initial settings will cause the long-term behaviour of the sequence to change beyond recognition. This is poetically known as the *butterfly effect*.

The equations that govern our weather are chaotic, just as the logistic map is. So, even a minuscule change in atmospheric conditions, such as that caused by a butterfly fluttering its wings on one side of the world, can totally change the eventual outcome on the other side of the world, years later. It can potentially make the difference between a tornado forming or not. This phenomenon is why weather forecasters are sometimes so unpopular: the long-term forecasting of the weather is impossible with any accuracy.

Chaotic star systems

The origins of chaos theory lie in astronomy. In space, solar systems consist of bodies such as stars and planets orbiting around each other under the mutual attraction of gravity. When there are two such bodies, for example a planet orbiting a star, the paths they follow are easy to predict. It was Johannes Kepler, in the 17th century, who first unlocked the secrets of planetary motion, proving that planets orbit the sun not in circles, typically, but in ellipses. In other parts of the galaxy, there are solar systems quite unlike our own, where twin stars are locked in orbit around each other.

Whenever just two objects are involved, it is mathematically easy to predict the paths that gravity will carry them along. Roughly speaking, if they are moving slowly, they will either spiral around each other getting closer and closer until they collide, or they will orbit each other indefinitely in stable orbits of interlocking ellipses. If their speeds are high, then they will fly apart along predictable paths called parabolas or hyperbolas, never to meet again. (This is what happens when single-apparition comets appear in the sky, just once before flying off forever. Comets that reappear, such as Halley's, have large elliptic orbits.)

'Float like a butterfly, sting like a bee.'

<small>Mohammed Ali</small>

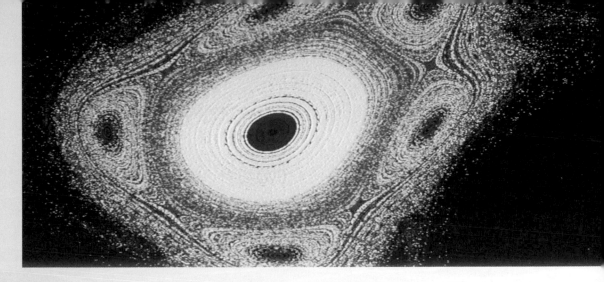

The moral is that gravitational systems comprising two objects are fairly easy to understand. When a third body is introduced, the situation becomes hugely more complex. In fact, it becomes chaotic. The paths that the three objects will follow are highly complex, and never repeat themselves, quite unlike the clean geometry of spirals, ellipses and parabolas. The slightest alteration in the initial configuration produces dramatically different results. This is known as the *three-body problem*, and it was the first discovered example of chaos in physics. As Isaac Newton put it, 'to define these motions by exact laws admitting of easy calculation exceeds, if I am not mistaken, the force of any human mind'.

A strange attractor

The strange attraction of strange attractors

In the logistic map on page 47, the input was a single number. More complex systems may instead have inputs consisting of pairs or triplets of numbers. In such systems, chaos is revealed in its most wondrous form.

The attracting cycles we saw in the logistic map were of simple kinds: either the sequence got closer and closer to a single point, or it jumped back and forth between two points, or between four, or eight, and so on. Chaotic systems on the 2-dimensional plane show a new type of behaviour, far more complex than this.

Whichever point on the plane we choose as the initial input, the system will pull it towards a particular region of the plane, known as its *attractor*. Often these attractors are not simple geometric shapes, but fabulously intricate and beautiful patterns. When these sets have *fractal dimension* not equal to a whole number (see *How to make a million on the stock market*), they are called *strange attractors*. For this reason, chaos theory is intimately bound up with the geometry of fractals.

8 How to survive a whirlpool

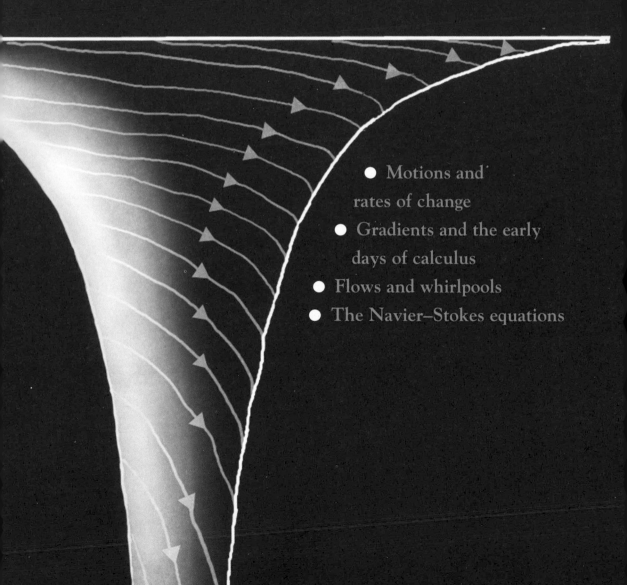

- Motions and rates of change
- Gradients and the early days of calculus
- Flows and whirlpools
- The Navier–Stokes equations

Change can be distressing for the human mind. It is a perennial problem, and this is true no less in the physical sciences than in any other walk of life. To use mathematics to describe a static system is one thing. To analyze a system that evolves over time is considerably more difficult. In fact, mathematics is superbly well suited to the analysis of change, but it took many centuries for the correct techniques to be found.

A picture is worth a thousand words

A simple example of change is a moving object, such as a train, travelling down its tracks. Here, the change is in terms of position. A train has the advantage that it travels in a fixed direction, so its position at any moment can be captured by a single number: its distance from the starting signal. (A fly buzzing around a room will need three coordinates to describe its position.)

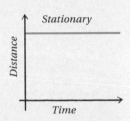

The train's situation can be expressed as a graph by plotting time on the horizontal axis, against its position on the vertical axis. This simple distance–time graph may not seem much, but it actually represents a huge step forward. The changing system has now been captured by something static: a single geometric curve.

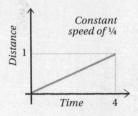

The graph describes the train's position at each moment, but there are other things we might need to know about it, chief among them is its speed. The simplest case is when the train is moving at constant speed. In this case, the graph will come out as a straight line. The faster the train is moving, the steeper this line will be. If the train is actually stationary, the line will be horizontal, indicating that its position is the same at every moment.

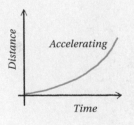

Although Archimedes never saw a train, he would have understood the relationship between the graph of position against time, and the train's speed. Speed is expressed by the steepness of the graph, or, as mathematicians call it, the graph's *gradient*. In general, gradients represent *rates of change*. In this case speed is the rate of change of position. This observation was to be one of the most important in the history of science.

Gradients: fighting an uphill battle

Archimedes knew the importance of rates of change. In the example of the train, it is position that changes. In other contexts, other quantities may change. During radioactive decay, the mass of an object changes with time. As music is played, the volume and pitch of sound both change.

If we are interested in some quantity y, and t represents time, then the rate of change of y is written as $\frac{dy}{dt}$ (pronounced 'dee-wye by dee-tee'), or sometimes $\frac{\partial y}{\partial t}$. In the example of the train, y is distance and $\frac{dy}{dt}$ is the rate of change of distance, better known as speed. The intuition is that $\frac{dy}{dt}$ should be a small change in y divided by the corresponding small change in t. Unfortunately, despite appearances, we usually cannot calculate $\frac{dy}{dt}$ just by dividing some number 'dy' by another 'dt'.

In mathematical terms the problem is the same in every case: given y, calculate $\frac{dy}{dt}$. Archimedes made the crucial observation that $\frac{dy}{dt}$ is the gradient of the graph of y. What was missing was the technique to calculate this exactly.

The idea of measuring a gradient is familiar to anyone who has ever driven a car up a hill. At the foot of the hill there is usually a road-sign, which might read 25%. What this means is that for every metre travelled horizontally, the road rises by a quarter of a metre vertically. This is the definition of the gradient: the amount of height gained per unit of horizontal distance travelled. Mathematicians generally represent it as a fraction rather than a percentage, so the central graph on page 51 has a gradient of $\frac{1}{4}$.

For a straight line the gradient is easy to calculate. Just as for a hill we calculate the height gained over some stretch, and divide it by the horizontal length. For distance–time graphs, the graph's height (y) represents the train's distance from the start, and the horizontal axis (t) represents the time it has been travelling. So the gradient $\frac{dy}{dt}$ gives the distance divided by the time, that is to say the speed. For curved graphs, the principle is the same, but the answer is more difficult to calculate.

● The artistic troubles of Archimedes

When we have a curved line instead of a straight one, the steepness varies from point to point. At some places it may be very steep, that is to say the gradient $\frac{dy}{dt}$ is a large number. At others, it may be totally flat, meaning, $\frac{dy}{dt} = 0$. To calculate the gradient at a particular point, Archimedes had to rely on a rather primitive method. First he drew a straight line that touches the curve just once, at the point of interest. This is called a *tangent* to the graph. Then he calculated the gradient of the tangent, as before.

This is a profoundly inadequate method, since it entirely relied on Archimedes' artistic ability. If he could sketch both the original curve and its tangent line with reasonable accuracy, he got a correspondingly accurate answer for $\frac{dy}{dt}$. If he messed up the drawing, his answer would be completely wrong.

This presented a serious obstacle to the applicability of mathematics to the wider world. Rates of change are vital to understanding many aspects of science, but mathematicians had no way to calculate them without drawing elaborate pictures, and risking hopeless inaccuracy. This gap was filled in the 17th century by two of the greatest thinkers of the age, Isaac Newton in Britain and Gottfried Leibniz in Germany.

A mathematical revolution: the dawn of calculus

Newton and Leibniz's idea was as follows: suppose we want to know the speed that a train is travelling at a particular instant. This had been impossible to calculate directly. What was feasible, however, was to gauge an approximation to this speed, by calculating its average speed over a larger range: for example, we could calculate the average speed over 10 seconds, simply by dividing the distance covered in that time by 10. Of course, if the object is not moving at constant speed, this will not equal exactly what we want. A more accurate estimate would be to use a smaller range of perhaps 1 second, or 0.1 second, or 0.01 second. What Newton and Leibniz realized was that, when we consider ever tinier ranges, the answer will zero in on the true value of $\frac{dy}{dt}$ at the point we want.

Tangent

Approximate tangents

The idea was brilliant. What is more, the algebraic rules that emerged were surprisingly easy to implement. The most important rule was as follows: if $y = t^2$, then $\frac{dy}{dt} = 2t$. If $y = t^3$, then $\frac{dy}{dt} = 3t^2$, and generally, if $y = t^n$ then $\frac{dy}{dt} = nt^{n-1}$.

Just by applying this rule, and a few others (none of them especially complex), the problem of artistic ineptitude evaporated. Rates of change could be calculated without drawing a single picture. This new technique of *calculus* was a triumph, arguably the most important moment in the history of mathematics. For the first time it brought the study of complex changing systems within the grasp of the mathematical sciences.

For Gottfried Leibniz, this accomplishment came at great personal cost. It became a matter of fierce contention whether the discovery should be attributed to Newton or to Leibniz. Newton was adamant that Leibniz had stolen his idea, and Newton had power on his side. He was the president of the Royal Society in London, in which capacity he arranged an enquiry to determine to whom the discovery belonged. Unsurprisingly (since he wrote it himself) the enquiry came down squarely on Newton's side, accusing Leibniz of plagiarizing his work, from which slur he never recovered. It was a needless and tragic outcome; the inclusion of both men among history's very greatest scientists is beyond dispute.

Calculus takes on the oceans and skies

The development of calculus allowed mathematicians to build a whole new armoury of tools for analyzing change in the wider world. One of the most important examples was the study of *flows*. Many things that are important to human beings flow: without the flow of blood we are nothing but a pile of soggy cells. Without the flow of money, we would be reduced to subsistence farming. But flows are notoriously difficult to understand, even with the new power of calculus. Strange phenomena occur, such as unpredictable and chaotic turbulence, or whirlpools unexpectedly appearing. The discovery of calculus at last provided a way in, and the study of fluid flows was begun in earnest by Leonhard Euler in 1757.

How can we model a mathematical object as a flow? When studying a train travelling down the track, the data to input was the time since the train started running. From that the graph told us its distance from the start, and then we could calculate its speed. In a flow, what we want to know is the direction and speed at which the fluid is flowing. This varies not just from moment to moment, but from place to place. If water is swishing about inside a fish-tank, in the corner of the tank the water may be nearly stationary, but in the middle it may be moving quickly. Similarly, in a whirlpool, the nearer the centre you go, the more tightly the fluid circulates.

A *fluid model* takes as inputs the spatial coordinates of a point and a moment in time, and returns as outputs the direction and speed of the fluid at that location, and at that moment. The central question is how this flow changes from moment to moment.

The laws of fluids

Leonhard Euler used the new mathematics of calculus to study the physics of fluids. For the first time, he wrote down an equation to describe how a flowing fluid evolves over time. However, Euler's work relied on two fundamental assumptions. First, he assumed that the fluid is *incompressible*. That is to say, it can be pushed around but can never expand or contract to fill a larger or smaller space. While this is not a totally accurate picture, many physical liquids are close to being incompressible in ordinary circumstances. Euler's second assumption was more problematic: he assumed the fluid was *inviscid*. This means that the fluid has no internal frictional forces that resist motion and slow it down. This assumption is flagrantly contradicted by many thick fluids, such as treacle. Even water has some viscosity (which is why doing aerobics in water is more energy-consuming than doing it on land). Meanwhile blood is more viscous than water, confirming the old adage, and making Euler's equation of little use to those studying the human circulatory system.

In the early 19th century two men, Claude-Louis Navier and Gabriel Stokes, set about extending Euler's analysis to more realistic, viscous fluids. The result of their work is the *Navier–Stokes equation*, which provides a description of the flow of viscous liquids. It looks, admittedly, rather alarming:

$$\frac{\partial \mathbf{u}}{\partial t} + (\mathbf{u} \cdot \nabla)\mathbf{u} = \nu \nabla^2 \mathbf{u} - \nabla p + \mathbf{f}$$

The central object described is the flow $\mathbf{u}$, whether this is blood around the body, or air from a burst balloon. The Navier–Stokes equation describes a condition imposed on $\frac{\partial \mathbf{u}}{\partial t}$, which is the rate at which the flow changes. The terms $(\mathbf{u} \cdot \nabla)\mathbf{u}$ and $\nabla^2\mathbf{u}$ are expressions of *vector calculus*, the flashy modern descendant of Newton and Leibniz's original calculus. These describe how the flow varies from point to point. The next term p incorporates the density and pressure of the fluid (which may also vary from place to place), and ν is a number that measures its viscosity (this is the term that Euler had missed out). The final term is $\mathbf{f}$, which describes the external forces on the fluid, the most important being gravity.

Flow wanted, any flow . . .

The principal aim of fluid mechanics for almost 200 years has been to solve this equation. Extraordinarily, this search has so far been unsuccessful. We can see innumerable types of flow around the world, from waterfalls to lava-lamps, and the work of Euler, Navier and Stokes tells us that every fluid flow must obey this law. All the same, mathematicians have as yet failed to find even a single example of a mathematical flow that does. In 2000, the Clay Mathematics Institute announced this as one of their Millennium problems, meaning that if anyone does find a flow that acts as we believe every flow must, they can claim $1,000,000 on top of the world-wide recognition they would thoroughly deserve.

9 How to make a million on the stock market

- Richardson's war
- How to measure a line
- Fractals and fractal dimension
- Mandelbrot and the
  fractal stock exchange

Lewis Fry Richardson is one of Britain's most underappreciated scientists. He made significant advances in a diverse range of sciences including weather forecasting, fluid dynamics and statistics. He also made major inroads into the study of certain intricate self-repeating patterns. Today these exotic shapes are known as fractals, but this term was coined long after Richardson's work initiated the subject.

● Richardson's war

War had a major impact on Richardson's life. Raised a Quaker, his pacifist beliefs remained with him throughout his life. He was a conscientious objector during the First World War. Not only did he refuse to fight, he declined to participate in any scientific research that could be put to military use. Instead he volunteered as a driver for the Friends' Ambulance Service on the front-line and gained a reputation as a conscientious driver, even under shell-fire.

One result of his refusal to join the military was that Richardson was subsequently disbarred from holding university positions. However, he continued with his wide-ranging scientific research at the Meteorological Office in London, and then at technical training colleges in England and Scotland. His work on predicting weather patterns was ground-breaking, and earned him election to the Royal Society in 1926.

War was also a continual theme in Richardson's work. He was particularly interested in the causes of war. Of course, every conflict has its own proximate causes, which historians study. The First World War, for example, was ultimately triggered by the assassination of Archduke Franz Ferdinand, following decades of manoeuvring by the European powers.

Over a longer timescale, there are more general phenomena at work. For example, minor skirmishes are much more common than all-out war. Richardson believed that statistical analysis could contribute a great deal to this discussion, and in his 1950 work *The Statistics of Deadly Quarrels* he compiled data for all conflicts since 1820 (where available), numbering 108 in total.

Richardson quantified the seriousness of each conflict by the number of dead, and set about comparing the frequency with which nations go to war against a large variety of other factors, from economics and religion to whether states are have sea-coasts or are landlocked.

War and peace and distance

One of the factors Richardson investigated was whether the likelihood of two countries going to war was influenced by the length of their common border. Do countries with short borders fight more or less than those with long borders? Are those conflicts likely to be minor or major? To investigate such questions, Richardson set about collecting data on the lengths of various international boundaries. However, here, he encountered an unexpected problem. The lengths of international borders are extremely ill-defined. For example, according to Spanish data, the border with Portugal is 987 km long. The Portuguese measure the same border at 1,214 km.

Was it a simple mistake? The phenomenon recurred too often, and the differences were too stark for that to be the case. So Richardson found himself having to go back to basics, and consider the problem of how to measure the length of a line.

How to measure a line

When a line is perfectly straight, of course there is no difficulty. Most international boundaries are not like this, however, and coastlines even less so. When a line is very wiggly, you have to make a choice about the scale you are going to use to measure it. To measure the coast of Great Britain, for example, you could start with a large-scale map and use a measurement with basic unit of length 100 km. This will give one answer. Using a more detailed map, and a basic length of 1 km you could produce a more accurate answer, which took into account more river estuaries and other wiggles. At the extreme, you could walk the entire coast of the country armed with a metre ruler.

It is no surprise that these methods will produce different answers. What you might expect is that these results should all be approximations to the true answer, of varying accuracy. For example, the first method might produce a result of 8,000 miles, the second a result of 10,000 miles and the third a result around 12,000 miles, with increasingly accurate measurements giving answers ever closer to the true answer of 12,213 miles (for example).

That is certainly what would happen if we adopted this approach to an ordinary geometrical shape such as a circle. By approximating it with straight lines of shorter and shorter lengths, we would eventually zero in on the exact length. But Richardson found that this does *not* happen with physical coastlines. He gathered data on various coastlines and boundaries, including the west coast of Great Britain, one of the most irregular coastlines in the world. In most cases he found that the measurements do not get closer and closer to some fixed number. Each time he zoomed in, he found many more kinks and wiggles to take into account, which significantly increased the final result. As the scale got shorter and shorter, the answers increased *without limit*. Today, this is known as the *Richardson effect*.

The emergence of fractals

The astonishing upshot of Richardson's analysis is that there simply is no meaningful number that can be assigned to the length of a coastline such as Britain's. The answer depends entirely on the scale used when taking the measurement. As Benoît Mandelbrot wrote in 1967, 'Geographical curves are so involved in their detail that their lengths are often infinite or, rather, undefinable.'

Mandelbrot's paper 'How long is the coast of Britain?' revived interest in Richardson's work. Mandelbrot is best known for his work on *fractals*, fantastically intricate geometric patterns, which are *self-similar*, meaning that they look the same however far you zoom in.

A simple example is the *Sierpiński carpet*, right. One of the first fractals, it was constructed by Wracław Sierpiński in the early 20th century. It has a comparatively straightforward construction: first draw a square. Then divide it up into nine smaller squares, and throw out the central one. Then divide each of the remaining eight squares into nine, and again throw out the middle one. Repeating this process *ad infinitum* will produce the Sierpiński carpet.

The defining characteristic of fractals is *self-similarity*, which comes from the following observation: if you zoom in on any patch of the pattern, what you see is an exact replica of the whole thing. A shrunken down version of the shape fits inside the original exactly. An interesting question is how many times it fits inside? This is addressed by the *fractal dimension* of the shape.

A journey into the 1.89th dimension

If we take a line 1 metre long, and shrink it to a third of its length, then the shorter line fits into the original three times. This much is obvious. However, if we start with a square, shrink it to a third of its length, then the shrunken square fits into the original nine times. Starting with a cube, and shrinking it to a third of its length, then the smaller cube fits into the original 27 times. What is the rule here, regarding dimension?

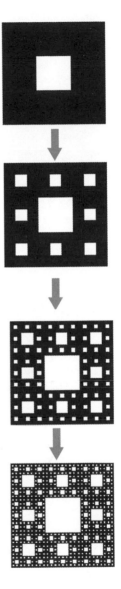

The first four stages of the Sierpiński carpet

For the 1-dimensional line, the smaller one fits in 3^1 times. For the 2-dimensional square, it fits in 3×3 times, that is to say 3^2. For the 3-dimensional cube, it fits in $3 \times 3 \times 3 = 3^3$ times. In each case, the dimension of the shape appears as a number of threes, or the *exponent*. So for a d-dimensional shape, the shrunken version fits into the original 3^d times.

Now what happens if we try to apply this to the Sierpiński carpet? If we shrink it to a third of its size, then the shrunken version fits into the original exactly 8 times. So if we want to measure its dimension as a number d, then it should be the case that $3^d = 8$.

Of course there is no whole number that satisfies this equation, but mathematicians can solve it nevertheless using *logarithms* (which are the opposite of exponents). The answer is that the Sierpiński carpet has a dimension of $\log_3 8 = 1.89$. Because this is between 1 and 2, the geometry of the carpet is between that of a line and a plane. This means that it is somehow dense; it fills up space more than an ordinary 1-dimensional line does, but not quite as much as a 2-dimensional square.

The first four stages of the Koch snowflake

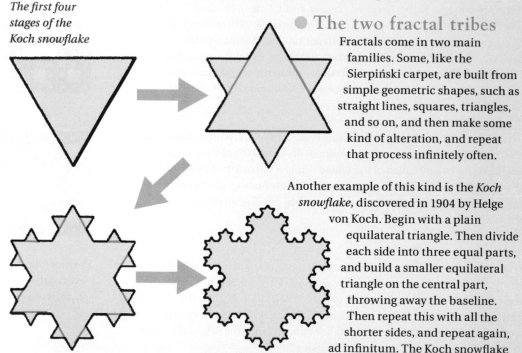

● The two fractal tribes

Fractals come in two main families. Some, like the Sierpiński carpet, are built from simple geometric shapes, such as straight lines, squares, triangles, and so on, and then make some kind of alteration, and repeat that process infinitely often.

Another example of this kind is the *Koch snowflake*, discovered in 1904 by Helge von Koch. Begin with a plain equilateral triangle. Then divide each side into three equal parts, and build a smaller equilateral triangle on the central part, throwing away the baseline. Then repeat this with all the shorter sides, and repeat again, ad infinitum. The Koch snowflake has a fractal dimension of $\log_3 4$, around 1.26. Just like a Richardson coastline, it is infinitely long: at each stage every line is divided into three parts and replaced with four parts of equal length, which entails increasing the overall length by a factor of $\frac{4}{3}$. Since this happens infinitely often, the total length grows larger and larger each time.

The second family of fractals are even subtler creatures, which have their origins in the mathematics of chaos theory. The most famous of these is the Mandelbrot set (below), which arises from a system similar to the logistic map (see *How to unleash chaos*).

● The strange geometry of money

Since its beginnings in Lewis Fry Richardson's work, fractal geometry has been a subject with applications to the real world. Physical coastlines have the property that whether you view them from near or afar, their level of wiggliness appears approximately the same, making them close approximations to fractals. Mandelbrot noticed another aspect of the human world with the same property: financial markets.

Mandelbrot assembled a comprehensive encyclopaedia of data on cotton prices going back for over a century. His 1963 paper analysing the results shocked the world of economics. Ordinary measures of data such as the mean or the variance were of little use, as they jumped around almost as much as the original data. Economists have a way to measure *volatility*; the problem was that the level of volatility itself was highly unstable, jumping around within a huge range.

How could such wild data possibly be tamed? Mandelbrot's game-changing observation was that the data looked essentially the same whether viewed on a timescale of minutes or years. It formed, in other words, a fractal. The consequences of this are profound, and continue to be investigated today. For one thing, it underlines the high-risk nature of commodity markets: occasional giant leaps in price, either up or down, are not freakish failures of the system, but inherent features of it.

> '*The idea that you can actually predict what's going to happen contradicts my way of looking at the market.*'
>
> GEORGE SOROS

The Mandelbrot set

10 How to outrun a speeding bullet

- Achilles and the tortoise
- Convergent and divergent series
- How to do infinitely many things in a short space of time
- Zeno's dichotomy paradox

It is a classic western scene: two gunfighters stand facing each other, Stetsons on, and trigger fingers at the ready. Then, in a surreal twist, one of them turns around, and slowly strolls away. The other fires at the retreating man's back, but somehow the bullet gets stuck in the air on the way to its target, slowing down as if travelling through treacle, until it reaches a standstill just before the point of contact.

It sounds like science fiction, but around 450 BC Zeno of Elea advanced an argument that would mean that the retiring gunman need have no concern at all, but could walk away safe in the knowledge that the bullet will never catch him. It would be tempting to dismiss Zeno's ideas as the ramblings of a lunatic, if it were not for the fact that his argument has a strangely compelling logic to it. In fact, Zeno was anticipating ideas that would not fully enter mathematics for over a thousand years.

Of course, Zeno did not have in mind a race between a bullet and a cowboy; rather, he envisaged a contest between the Greek hero Achilles, famed for his strength and speed, and a humble tortoise, with a (rarely achieved) top speed of 0.3 miles per hour.

The unfair race

Mathematics certainly contains many surprising and counterintuitive facts. Georg Cantor's work on infinity is a striking example (see *How to count to infinity*). But Zeno's argument really takes this to its limit. He argued that the great Achilles would never be able to defeat the feeble tortoise in a race.

The argument goes as follows: suppose Achilles is to race the tortoise over 10 metres, and agrees to give the creature a head start, of say 5 metres. To win the race, Achilles needs to cross the finish line. Before he does that, he needs to reach the half-way point, which is where the tortoise begins.

Here is the problem: by the time Achilles reaches the place where the tortoise started, a little time has elapsed and so the animal has moved on by a small amount. So Achilles must now run to the tortoise's new position. By the time he does this, though, it has moved on a tiny bit more. This process keeps happening: whenever Achilles reaches the point where the tortoise formerly stood, a little time has passed, and the tortoise has moved forwards a tiny bit more. However fast Achilles runs, he is always playing catch-up; the tortoise will always remain one step ahead. Exactly the same argument applies to the bullet heading towards the retiring gunslinger.

● Zeno's mathematical philosophy

We do not know a great deal about Zeno, as none of his works survive, but come to be used indirectly via Plato and Aristotle. It seems, however, that he was a philosopher of truly radical beliefs. He held that motion, and indeed any form of change, is an illusion. In support of this view, he compiled a list of paradoxes, of which Achilles and the tortoise is the most famous. Zeno also believed that it was impossible to subdivide the world into smaller components such as earth and sky, or me and you; all such divisions too are illusory. All that exists, according to Zeno, is a single indivisible entity.

This mystical view may still have an appeal to some people of a religious disposition. Not many, however, would attempt a mathematical defence of it, as Zeno did. However, his paradoxes have an importance that transcends his metaphysical beliefs, since they anticipated mathematical discoveries that would not be made for over a thousand years.

● How to do infinitely many things in a short space of time

Not to put too fine a point upon it, Zeno was wrong: great athletes can outrun tortoises, and a walking person is not safe from a pursuing bullet. This should not come as a major surprise. At any rate, I would not wish to encourage anyone to test this hypothesis. The important question, then, is how to overcome Zeno's objection, that Achilles has an infinite number of tasks to complete before he can overtake the tortoise.

It would take infinitely long to list all the tasks that Achilles needs to complete: first he must reach the tortoise's initial position. Then he must reach its second, then its third, then its fourth, fifth, and so on.

If we wanted to list everything that Achilles needs to do, we would still be working on it at the end of the universe. That is what makes the problem so compelling. However, just because it would take infinitely long to *say* the tasks that Achilles must complete, it does not follow that it must take him infinitely long to *do* them. This is the hidden false step in the argument.

Suppose that it takes Achilles 1 second to reach the tortoise's starting position, and then it takes him $\frac{1}{2}$ a second to run from there to the tortoise's next position, a further $\frac{1}{4}$ of a second to run to its third position, and so on. (The numbers here are chosen to illustrate the central point, rather than accurately represent the speeds of the racers.) The total time that it will take him to catch the animal is therefore: $1 + \frac{1}{2} + \frac{1}{4} + \frac{1}{8} + \frac{1}{16} + \cdots$

This is what mathematicians call a *series*: a list of terms that are added up as we progress along. This particular series has an important property. Even though there are infinitely many terms in the series, as we move along them the total does not get bigger and bigger without limit. Adding up the first four terms gives $1\frac{7}{8}$. Adding up the first ten gives $1\frac{511}{512}$. As we add up more and more terms, we never reach any huge numbers, instead we get closer and closer to 2. The mathematical way of saying this is that the series *converges* to the value 2. We might write this as:

$$1 + \frac{1}{2} + \frac{1}{4} + \frac{1}{8} + \frac{1}{16} + \cdots = 2$$

Back at the race, after 2 seconds, all of Achilles' intermediate steps have been completed. At that moment he will overtake the tortoise. The moral is that it *is* possible to carry out an infinite number of tasks within a finite amount of time, so long as the times for each task form a convergent series.

The eternal mantra

When we try to list the tasks Achilles has to do, we end up repeating a single phrase like a meditative mantra: 'then he has to run to its next position, then he has to run to its next position, then he has to run to its next position . . .'. Let's suppose that it takes 1 second to say this phrase once. Then the time it will take to list all his tasks can again be expressed as a series: $1 + 1 + 1 + 1 + 1 + \cdots$

Unlike the previous series, this does not converge to a fixed value. It just grows ever bigger, without limit. Such a series is called *divergent*. It is because this series is divergent that Zeno's argument seems convincing: the act of listing Achilles' necessary tasks can indeed never be completed.

There are many series that are neither convergent nor divergent but jump around forever. One example is $1 - 1 + 1 - 1 + 1 - 1 + 1 - \cdots$, which fluctuates forever between 1 and 0. Another is $1 - 2 + 4 - 8 + 16 - 32 + \cdots$

Growing bigger, very slowly

More than a thousand years after Zeno described his race, series would come to occupy a central position in mathematics. The question of whether a particular series is convergent like $1 + \frac{1}{2} + \frac{1}{4} + \frac{1}{8} + \frac{1}{16} + \cdots$, or divergent like $1 + 1 + 1 + 1 + 1 + \cdots$, or neither, is one that is of enormous importance in other areas of mathematics. In these particular examples it is not too hard to tell. In other cases, it can be much more difficult.

A case in point is the *harmonic series*: $1 + \frac{1}{2} + \frac{1}{3} + \frac{1}{4} + \frac{1}{5} + \frac{1}{6} + \cdots$ Does this series converge or diverge? Successive terms certainly get smaller and smaller, which is the first requirement for convergence. If we add up the first 10 terms, we get an answer of just under 3, adding up the first 100 gives an answer of just over 5. The first 1,000 produce a result of around 7.5. So its rate of growth is slowing down, and it would be very tempting to assume that the result is converging to some fixed number.

Appearances can be deceptive though, in mathematics as much as in life. Around 1350, Nicole d'Oresme realized that this series actually diverges.

He reasoned as follows: after the initial 1, break up the series into blocks of length 1, 2, 4, 8, 16, and so on:

$$1 + \left(\frac{1}{2}\right) + \left(\frac{1}{3} + \frac{1}{4}\right) + \left(\frac{1}{5} + \frac{1}{6} + \frac{1}{7} + \frac{1}{8}\right) + \cdots$$

Now each bracket has a value of at least $\frac{1}{2}$. For example, $\frac{1}{4} + \frac{1}{4}$ equals $\frac{1}{2}$, and $\frac{1}{3} + \frac{1}{4}$ is a little bit more. Similarly, four lots of $\frac{1}{8}$ come to $\frac{1}{2}$, and $\frac{1}{5} + \frac{1}{6} + \frac{1}{7} + \frac{1}{8}$ is slightly more than this. So, by reducing the series slightly, we get a new series $\frac{1}{2} + \frac{1}{2} + \frac{1}{2} + \frac{1}{2} \ldots$ Of course this diverges, and so it follows that the original harmonic series must do.

When this fact was first discovered, it was a shocking result. The reason is that although the harmonic series diverges, it does so exceptionally slowly. The fact of divergence means that the series will eventually surpass any number you care to name, a million, a billion, anything. However to reach a total of even a hundred, we would need to add together the first 15,092,688,622,113,788,323,693,563,264,538,101,449,859,497 terms of the series!

Achilles at a standstill

Convergent and divergent series are not the only elements of modern analysis hidden within Zeno's work. In a second paradox, he advanced an even more outrageous argument. He claimed that, even if Achilles was in training on his own, with no tortoise for miles around, he would still not be able to complete his run.

The argument went as follows: to complete the race, Achilles first needs to reach the half-way mark. But before he can do that, he needs to reach the quarter-way mark. Before that, he needs to reach the one-eighth mark, and so on. This time there is not even a first step that Achilles can take without getting caught in this infinite regress. So he can never even get started.

The same argument would imply that a gunslinger aiming at a barn door at close range will not be able to hit it. The bullet will never even leave the barrel of his gun: for it to do so, it first has to reach the half-way point, and so on, ad infinitum. For Zeno, this was proof positive of the illusory nature of motion. However, the idea of understanding motion through the analysis of infinite sequences of decreasing lengths would later become one of the most important developments in mathematical history. When Isaac Newton and Gottfried Leibnitz picked up this idea in the 17th century, it would come to be known as *calculus* (see *How to survive a whirlpool*).

11 How to solve the Da Vinci code

- Fibonacci's rabbit experiment
- The Fibonacci sequence
- The golden section in art and nature
- Bernoulli's golden spiral

The rate at which rabbits reproduce is the stuff of legend. But how fast do they actually breed? In 1202, Leonardo da Pisa used his mathematical skills to try to solve the problem more precisely. Better known by the name of Fibonacci, he could little have expected that his analysis of rabbit-breeding would become one of the most famous pieces of mathematics in history.

Fibonacci had travelled around the Islamic world, studying algebra, and it was he who popularized the system of Arabic numerals in Christian Europe, noticing the superiority of expressions such as '48' to the clunky old Roman numerals XXXXVIII. This was crucial groundwork for the coming scientific renaissance. However, his name would forever be associated with his thought experiment about rabbits: the *Fibonacci sequence*, as it became known. Through this sequence, he would also reveal another wonder of the mathematical world, the celebrated *golden section*. These two would become emblematic of how hidden patterns among numbers can have an uncanny ability to describe unexpected aspects of the world.

The reproduction of rabbits

The question Fibonacci set out to answer was this: suppose a pair of rabbits is let loose in a large garden, surrounded by walls, and containing no predators. As rabbits do, they will start breeding. The question is: how many rabbits will there be in a year's time, or in two years, or ten?

To study the rabbit problem, Fibonacci first had to build a theoretical model of it. Being a mathematician rather than an expert in rabbits, he made several rather crude assumptions. Firstly he took *pairs* of rabbits as the fundamental unit to count, assuming that pairs are born together and remain monogamous for life. Baby rabbits, he assumed cannot reproduce, but mature into adults after 1 month. Each month every pair of adult rabbits produces a new pair of babies. Finally, he assumed that rabbits never die. These assumptions took Fibonacci some distance away from a realistic model of rabbit reproduction. However, the simplicity and elegance of the resulting system lends it great importance in a range of scientific scenarios.

Fibonacci had set the rules for his abstract rabbit-system, so what happens when we start it and let it run?

We begin by introducing one pair of baby rabbits into the garden. The next month, they have matured into one pair of adults. The following month they have reproduced, so we have the original pair of adult rabbits together with a new pair of baby rabbits. The month after that, these baby rabbits have matured, and the original adults have reproduced again. So we have two pairs of adult rabbits, plus one pair of babies, and so on. If we count the number of pairs each month, we obtain a sequence: 1, 1, 2, 3, 5, 8, 13, 21, 34, 55, 89, 144, . . .

● Sunflowers and pineapples

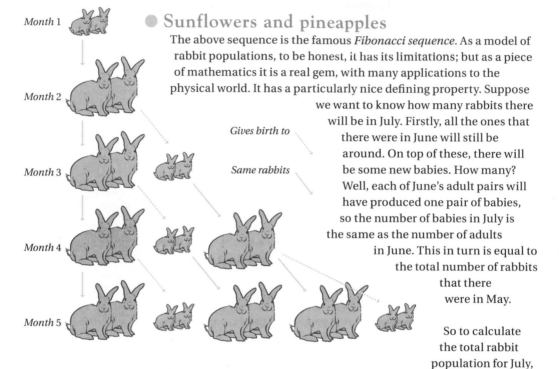

Month 1

Month 2

Gives birth to

Month 3

Same rabbits

Month 4

Month 5

The above sequence is the famous *Fibonacci sequence*. As a model of rabbit populations, to be honest, it has its limitations; but as a piece of mathematics it is a real gem, with many applications to the physical world. It has a particularly nice defining property. Suppose we want to know how many rabbits there will be in July. Firstly, all the ones that there were in June will still be around. On top of these, there will be some new babies. How many? Well, each of June's adult pairs will have produced one pair of babies, so the number of babies in July is the same as the number of adults in June. This in turn is equal to the total number of rabbits that there were in May.

So to calculate the total rabbit population for July, we need to add together the total numbers for May and June. This pattern can be seen clearly in the Fibonacci sequence: each term is the sum of the preceding two: 1 + 1 = 2, 1 + 2 = 3, 2 + 3 = 5, 3 + 5 = 8, and so on.

Fibonacci numbers are simply the numbers that appear in the Fibonacci sequence. They have a remarkable tendency to appear in surprising places in the natural world. The number of petals on many types of flower is often a Fibonacci number, for example. Similarly, if you look at a pineapple, the little fruitlets are arranged in spirals around the fruit. If you count these spirals, as often as not the number is a Fibonacci number. The same goes for the head of a sunflower. Why should this be? The answer is to do with the Fibonacci sequence's relationship to two other mathematical treasures: the golden section, and the logarithmic spiral.

The golden section: from Euclid to Leonardo da Vinci

Take a line, 1 metre long. The aim is to divide it into two parts, so that the ratio of the long part to the short part is the same as the ratio of the whole line to the longer part. This question was considered by Euclid himself around 300 BC. The number that accomplishes this is today called the *golden section*, and is usually represented by the Greek letter phi, ϕ. Its exact value is $\frac{1+\sqrt{5}}{2}$, which is around 1.618. Being an irrational number (see *How to become a celebrity mathematician*), its decimal representation never ends or repeats itself however. There are many interesting things about the golden section. For example, it has another beautiful representation, as:

$$\phi = 1 + \cfrac{1}{1 + \cfrac{1}{1 + \cfrac{1}{1 + \cdots}}}$$

This is an example of what mathematicians call a *continued fraction*.

The golden section

a $\qquad$ b $\qquad$ $\frac{a+b}{a}$ is the same as $\frac{a}{b}$

A *golden rectangle* is one whose sides are in proportion given by ϕ. It has another definition too: if you remove a square built from the shorter side, the rectangle that remains has the same proportions as the original. It is often said that a golden rectangle is the shape that is most pleasing to the eye. Certainly, the golden section has been a source of inspiration to mathematically inclined artists and architects for centuries, including Leonardo da Vinci himself. More controversially, it is sometimes claimed that the great landmarks of the past, from the Egyptian pyramids to Notre-Dame Cathedral, also have the golden section written into their designs. Every such claim must be judged on its merits, of course, but some scientific scepticism is required here. Just as King Midas turned everything he touched to gold, some people today have a tendency to see the golden section everywhere they look.

Why rabbits are golden

It is not immediately obvious that the Fibonacci sequence and the golden section should be related. To see the connection, we start by looking at the ratios of successive Fibonacci numbers: $\frac{1}{1}, \frac{2}{1}, \frac{3}{2}, \frac{5}{3}, \frac{8}{5}, \frac{13}{8}, \ldots$. As this sequence continues, it gets ever closer to a fixed number. A little algebra is enough to reveal this limiting number to be ϕ. By the time we get to $\frac{144}{89}$, this is already a reasonable approximation to ϕ, accurate to three decimal places. Of course the sequence never actually reaches ϕ, but it does get as close as you like.

The golden section is also the key to solving Fibonacci's original rabbit problem. Suppose he wanted to know how many rabbits there would be after 100 months. One method would be to write out the Fibonacci sequence, until we reach the hundredth term. It would be more convenient however to have a single formula that can calculate it directly. Such a formula does exist, and the golden section is at its centre:

$$F_n = \frac{\phi^n - (1-\phi)^{-n}}{\sqrt{5}}$$

This is known as *Binet's formula* after Jacques Binet who deduced it in 1843 (although it was already known to several earlier mathematicians, including Leonhard Euler). Now, without working through 99 intermediate numbers we can tell immediately that the hundredth Fibonacci number is $\frac{\phi^{100} - (1-\phi)^{-100}}{\sqrt{5}}$, which comes out as 354,224,848,179,261,915,075 pairs of rabbits.

A galaxy of spirals

Start with a piece of squared paper and draw a 1×1 square in the middle of the page. Next to it draw another. Together these form a 2×1 rectangle. On the longer side, draw a 2×2 square. Altogether we now have a 3×2 rectangle. Along the longer side of this, draw a 3×3 square. Altogether, this produces a 5×3 rectangle, along the side of which we draw 5×5 square. We continue like this, until we run out of space on the page. What results is a pattern of squares spiralling outwards. By connecting their corners with curves, we can extract a spiral, called the *Fibonacci spiral*. The connection, in case you missed it, is that sides of the squares are given by the Fibonacci sequence.

The Fibonacci spiral is a good approximation to the *golden spiral*, illustrated. Instead of squares, if we begin with a golden rectangle, and take out the square of the shorter side, we are left with another golden rectangle. If we repeat the process, we get another, and so on. Continuing like this, we obtain an ever-decreasing sequence of golden rectangles. If we join the corners of these, we get what is called a golden spiral. Unlike the Fibonacci spiral, this gets ever more tightly coiled as it spirals in forever.

Jacob Bernoulli later dubbed this *Spira Mirabilis*, 'the miraculous spiral'. He discovered it through a different route. If you draw a straight line from the centre to any point on the spiral, the angle between the line and the curve is

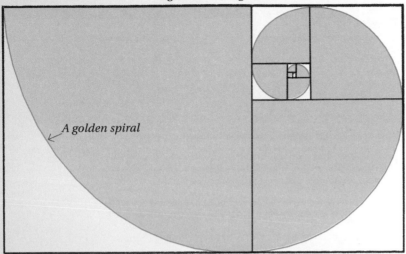

A golden rectangle

A golden spiral

always the same (technically $\frac{2\ln\phi}{\pi}$). Bernoulli was amazed by this spiral. Firstly, if you shrink or expand the whole thing by a certain factor (in fact 2π), the resulting shape remains exactly the same. This makes the golden spiral an early example of a *fractal* (see *How to make a million on the stock market*). Secondly even if you take the *inverse* of the spiral (a technical way to flip it inside out) the resulting curve is still the same. Not even most fractals have this property. Bernoulli was so in love with his spiral that he even asked that one should be engraved on his tombstone.

The golden spiral is the first in a family of *logarithmic spirals*, which often occur in the natural world, for example in the formation of spiral galaxies. The only difference is that the defining angle may be different from that of the golden spiral. This relationship is one reason for the Fibonacci sequence's propensity to appear in nature: it is the best way to approximate a logarithmic spiral, using only whole numbers.

The Da Vinci code

With its connection to so many mathematical treasures, and its tendency to appear unexpectedly in the natural world, it is not surprising that the Fibonacci sequence has achieved a celebrity surpassed only by the number π (see *How to square a circle*). One recent example of this was in Dan Brown's conspiracy theoretic novel *The Da Vinci Code*. The eponymous code begins with the numbers: *13 – 3 – 2 – 21 – 10 – 1 – 8 – 5*

The numbers are, of course, the first eight terms of the Fibonacci sequence, in the wrong order.

12 How to admire a mathematical masterpiece

- Euler's formula
- A mathematical trinity: addition, multiplication and exponentiation
- The amazing number e
- Euler's compass of the complex world

What does it mean to say that a piece of mathematics is beautiful? Concision is important: a good theorem should be quick to state, carrying a lot of information in a small amount of space. Yet it should not seem dense or cluttered. Like a good painting, it should be immediately striking to the reader, but without being gaudy or overly ostentatious. Like an expert ice-skater, it should seem natural and graceful, detached from the formidable technicalities that we know lie behind the display. Perhaps most importantly, like a good play, it should hold one's attention, and tell an interesting story, which elegantly interweaves familiar ideas in a new and unexpected way.

By near universal acclaim, the most beautiful of all theorems is *Euler's formula*, discovered by one of the greatest mathematicians of all time, the 18th-century Swiss virtuoso Leonhard Euler. Euler's mastery of mathematics was supreme; original ideas and great theorems simply poured out of him. By the end of his life, he had produced more research than anyone else in the subject's history, much of it of the absolute highest quality. He was an instrumental figure in many subjects such as graph theory (see *How to visit a hundred cities in one day*) and fluid dynamics (see *How to survive a whirlpool*).

His most famous contribution, however, was a short equation that gives a glimpse of the profound depths of the recently discovered world of complex numbers. The cast for Euler's formula is luxurious in the extreme, consisting of quite simply the five most important numbers there are, and telling the tale of a beautiful and subtle relationship between them. This relationship is expressed through the three most important mathematical operations: addition, multiplication and exponentiation.

Without further ado, here it is:

$$e^{i\pi} + 1 = 0$$

● An all-star cast

Of the five stars of Euler's formula, 0 and 1 need little introduction. Their significance is not in terms of what they do, but what they do not: when you add 0 to any number, that number is left completely unchanged. Meanwhile 1 plays the same role in respect of multiplication.

The next performer is more enigmatic, the number *i*. This is the foundation on which the new system of complex numbers was built (see *How to solve every equation there has ever been*). The number i is defined as $\sqrt{-1}$, meaning that $i \times i = -1$, and all other complex numbers can then be described using i.

A mathematical trinity: addition, multiplication and exponentiation

The leading role in Euler's formula is taken by the number e. For all their fame, the other four numbers form the supporting cast. In more technical terms, Euler's formula is a statement about the process of *exponentiation*.

For plain, whole numbers, exponentiation is not too complicated an idea. We know that multiplication is essentially repeated addition: $5 \times 3 = 5 + 5 + 5$ and $7 \times 4 = 7 + 7 + 7 + 7$. Similarly, exponentiation is multiplication repeated. So $5^3 = 5 \times 5 \times 5$, and $7^4 = 7 \times 7 \times 7 \times 7$.

This definition works perfectly well for whole numbers, but it is inadequate when we come to the wider world of numbers that mathematicians enjoy investigating. Particularly important are the complex numbers. What might 5^i mean, for example?

It is not obvious that 5^i need mean anything at all. Not every grammatically correct English sentence carries a sensible meaning, and the same is true for mathematical expressions. In fact, there are many meaningless expressions, popular examples being the likes of $\frac{7}{0}$ or $\frac{0}{0}$. Is complex exponentiation like these, pieces of mathematical gibberish?

As it turns out, it is not. It is possible to assign a meaning to things like 5^i and do so in a way that is perfectly consistent with the common interpretation of exponentiation, and indeed a very useful extension of it. This was one of the early triumphs of work into the complex numbers, in which Euler played a decisive part. The answer is given by a wonder of the mathematical world, called the *exponential function*. The public face of this object is the number e.

An aside on money

Suppose a bank account pays 10% interest per year. If I put £100 in at the beginning of the year, this will increase to £110 at the end of the year. At the end of the next year, what will the account contain? A common mistake is to think that it will increase by £10 per year, giving £120 at the end of the second year. In fact this is wrong; the account will increase by 10% of £110 in the second year. Of course this is £11, giving a total of £121. Suppose I want to know how much the account will contain in 20 years' time. It would be convenient not to have to work through 19 intermediate calculations.

How to admire a mathematical masterpiece

Well, each year, the amount in the account gets multiplied by 1.1. This is what increasing by 10% amounts to: 1, which represents the amount currently in the account, and another 0.1, which gives the 10% increase. So after 20 years we need to multiply by 1.1 twenty times. Exponentiation gives a convenient way to write this: $1.1^{20} \times 100$. This comes out as £672.75.

The bank account that pays all the time

It was noticed early on that the theory of compound interest has the number e lurking beneath it. To see this, consider a bank account that pays 100% interest per year. The catch is that you are only allowed to put £1 in it. After one year, the account will contain £2. Suppose instead though, that the account pays 50% every half-year. So, after one year, the amount in the account will be $1.5^2 \times 1 = £2.25$. What if it pays $33\frac{1}{3}\%$ three times a year? Then after one year, it will contain $\left(1\frac{1}{3}\right)^3 \times$ £1 $=$ £2.37, to the nearest penny. If we continue this trend, the next number will be $\left(1\frac{1}{4}\right)^4 = £2.44$, then $\left(1\frac{1}{5}\right)^5 = £2.49$, to the nearest penny, and so on.

What happens as we persevere with this line of thought? Will the numbers get ever larger, or will they approach some limit? In mathematical terms, the question is what happens to the sequence defined by $\left(1 + \frac{1}{n}\right)^n$ as n gets bigger and bigger? The answer is that it gets ever closer to a fixed number, known as e, which has a value of around 2.718282. So an account that provides interest *continuously*, meaning that it is growing at every single moment according to this rule, will end up with £e after one year.

The star of the show: the number e

This limit of the process of continuous interest is one definition of the number e. Another equivalent definition is:

$$e = 1 + 1 + \frac{1}{2} + \frac{1}{6} + \frac{1}{24} + \frac{1}{120} + \cdots$$

The pattern here is that $6 = 3 \times 2 \times 1$ and $24 = 4 \times 3 \times 2 \times 1$, and so on. These are written as 3! and 4!, where '!' is pronounced 'factorial' (see *How to count like a supercomputer*). So the standard way of writing e is as:

$$\frac{1}{0!} + \frac{1}{1!} + \frac{1}{2!} + \frac{1}{3!} + \frac{1}{4!} + \frac{1}{5!} + \cdots$$

(The first term may look a little strange, but it is conventional to take $0! = 1$.)

The power behind the throne

The number e is certainly important. Indeed there is a case that it is the most important number of all, from a mathematician's perspective. But it derives all its power from something larger. The power behind the throne is an object called the exponential function. This is like a machine that takes as input a number x and gives as output:

$$\frac{1}{0!} + \frac{x}{1!} + \frac{x^2}{2!} + \frac{x^3}{3!} + \frac{x^4}{4!} + \frac{x^5}{5!} + \cdots$$

To save writing this out every time, the exponential function is instead written as e^x. Both the name of the exponential function and the notation e^x are very suggestive of exponentiation. This is because the exponential function is indeed the right tool to lift exponentiation into the world of the complex numbers.

Euler's complex compass

The final number featuring in Euler's formula has achieved an international celebrity outstripping even that of i and e. It is a megastar of the mathematical world, the number π. This has its origins thousands of years ago, in the study of circles (see *How to square a circle*). In particular if a circle has radius r, then its circumference is $2\pi r$.

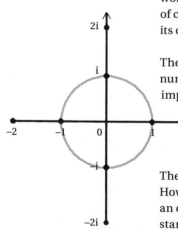

The circle Euler considered is one right in the middle of the complex numbers. It has its centre at 0 and has a radius of 1. This circle is a very important tool in understanding the world of complex numbers. As Euler proved, it acts as a sort of compass for the complex world. The directions east and west are marked by the numbers 1 and −1, while north and south are marked by i and $−i$ respectively.

The interesting action happens when we look at directions in between these. How can we measure intermediate bearings? The answer is the same as for an ordinary compass, by specifying the angle. While ordinary compasses start at north, and measure the angle clockwise, in mathematics we begin at the number 1 and proceed anticlockwise. There is a nice easy way to measure an angle we have reached: it is given by the distance travelled around the edge of the circle. The full distance around the edge is 2π by the ancient formula for the circumference of a circle (since this one has radius 1).

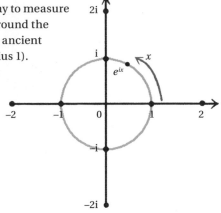

Dividing this up, a quarter turn amounts to a distance of $\frac{\pi}{2}$ and this will takes us as far as i. Similarly a three-quarter turn takes us to $-i$, and requires a distance of $\frac{3\pi}{2}$. The key observation for Euler's formula, is that a half-turn requires a distance of π and takes us to -1. (Known as *radians*, mathematicians much prefer this scale for measuring angles to the more common unit of degrees.)

The greatest advertisement

A compass is not just a disc with bearings written on it. At its centre is an ingenious piece of technology on which the whole thing depends. So it is with Euler's complex compass. His deep analysis of the complex numbers revealed something wonderfully surprising: the exponential function can be used to measure angles. This is the needle at the centre of the compass. More precisely, if you travel a distance of x around the central circle, the complex number you arrive at is e^{ix}. This was the fundamental theorem that Euler proved, and with it he provided a piece of equipment that would transform our knowledge of the complex world, and thereby the whole of mathematics and beyond. Quantum mechanics, for example, relies on a sophisticated understanding of the complex numbers (see *How to be alive and dead at the same time*).

Now every complex number could be captured by multiplying together two pieces of information: its bearing as expressed by the exponential function, e^{ix}, together with its distance, say r, from 0.

Euler's formula is ultimately a beautiful piece of advertising to demonstrate the capabilities of his new product.

We know that if you start at 1, and travel distance π around the central circle, you get to -1. So according to Euler's theorem, it must be true that $e^{i\pi} = -1$. Then only one final piece of rearrangement is needed to arrive at the mathematical equivalent to the Mona Lisa:

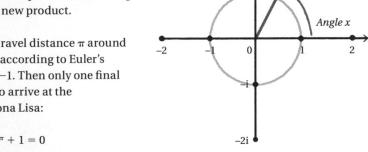

$$e^{i\pi} + 1 = 0$$

13 How to count like a supercomputer

- Fugues and factorials
- Permutations and combinations
- The birthday problem
- Partitions and the Hardy–Ramanujan formula

25×24
23×22
21×20
19×18

$25! \div 17!$

$n! \div (n-r)!$

$25 - 8 = 17$

The oldest mathematical artefact we have is the Lebombo bone. Dating from around 35000 BC, it is a baboon's leg bone, into which someone has carefully carved 29 notches. Some experts believed it was intended as a lunar calendar; whether or not this is true, it is a certainty that this tool was used for counting something. Counting is truly the most elementary and ancient form of mathematics. Indeed, it is the only branch of mathematics that we can say for certain extends beyond our species to any serious extent. Animals such as rhesus monkeys have a surprisingly good grasp of it, and are able to manipulate small numbers with skill rivalling that of humans.

You might have expected that in the millennia since the Lebombo bone was carved the theory of counting should be well and truly settled. Not so: under the guise of *combinatorics*, it still contains many questions and mysteries.

Counting fugues

A musical *fugue* begins with one instrument alone, playing the tune. After a little time, the next instrument joins in, playing the same tune, and then the third, and so on. Many fugues have been composed over the centuries, but the master of the form was the 18th-century German composer Johann Sebastian Bach.

Suppose our band comprises an accordion, some bagpipes and a clavichord (A, B and C). How many different orderings are possible? Because the numbers are small, we can count the possibilities directly. The first is ABC, meaning that the accordion plays first, then the bagpipes and then the clavichord. The other possibilities are ACB, BAC, BCA, CAB and CBA, giving a total of six. If we introduce an extra instrument, a didgeridoo (D), this will be a longer procedure. By the time we reach seven instruments the number of possible orderings is in the thousands. So mathematicians needed to find a short cut.

If we return to our three original instruments, for the first instrument to come in, there are three choices: A, B or C. For the second, one instrument has already been used, so there are two remaining possibilities. For the final instrument there is only one left. This suggests that the answer should be $3 \times 2 \times 1$. Happily, this agrees with the direct calculation above.

This also shows how to extend the problem to larger orchestras. For four instruments, there are four choices for the first one, and then three, then two,

and then one. This gives a total number of orderings of $4 \times 3 \times 2 \times 1 = 24$.
For seven instruments, we get $7 \times 6 \times 5 \times 4 \times 3 \times 2 \times 1 = 5{,}040$. This is what
mathematicians call the *factorial* function, written as '7!'.

It is not only for composers of fugues that factorials are important. They
appear throughout science wherever counting occurs, from analyzing DNA
bases to designing computers.

Factorials are the first in a long line of increasingly sophisticated techniques
for counting. The next are called *permutations*. Suppose that we want just
two instruments to play, in order, from our group of four. The choices are:
AB, BA, AC, CA, AD, DA, BC, CB, BD, DB, CD and DC.

Counting these gives 12 possibilities, but what is the underlying pattern
here? The reasoning begins in the same manner as factorials. For the first
instrument, there are four choices: A, B, C or D. Then for the second, as one
has been used, there are three remaining options. This gives a result of
$4 \times 3 = 12$, luckily matching with the above count.

We can extend this reasoning to situations where counting directly would be
impossible. So the number of orderings of 8 objects from 25 would be
$25 \times 24 \times 23 \times 22 \times 21 \times 20 \times 19 \times 18$, which is around 44 billion. These
different orderings are called *permutations*. A formula for this number, in
terms of factorials, is $25!/17!$ The 17 appears because $25 - 8 = 17$. So in
general, the number of orderings of r different instruments from a band of n
is $n \times (n - 1) \times (n - 2) \times \ldots \times (n - r + 1)$, which can be expressed concisely
as $n!/(n - r)!$

How to count like a supercomputer

How many flavours are there?

Biologists today believe that there are five basic tastes that our tongues can identify. The first four, saltiness, sweetness, bitterness and sourness, have been recognized for centuries. More recently, another basic taste has been recognized: savouriness, or *umami* (coming from the Japanese for 'tasty').

An interesting question is how many possible combinations can be made from these five. In fact, this question was considered in the first millennium BC, by the Indian medic Sushruta. He considered that there were six basic tastes: sweetness, sourness, saltiness, bitterness, astringency and pungency. He numbered the total combinations of these six as 63.

The simplest flavours come from the five tastes individually, giving five possibilities. Then we can consider tastes in pairs, such as sweet and sour. We can start by counting pairs of tastes just as for the musical instruments above. The number of permutations of two tastes from five is $5 \times 4 = 20$.

However, we have to be careful here, as it is not the number of permutations that we really want. For combinations of tastes, unlike for fugues, the order does not matter. Sweet and sour is the same as sour and sweet, and we do not want to double count. At the moment we have counted everything twice, so we need to divide by 2. So the number of taste pairings is $(5 \times 4)/2 = 10$. In general, the number of *combinations* of r different tastes from a total collection of n is $n!/(r! \times (n - r)!)$, where combinations unlike permutations do not care about the order.

We can use this formula to work out the number of combinations of three tastes from our five. It is $5!/(3! \times 2!) = 10$ again. Similarly the number of combinations of four tastes from five is $5!/(4! \times 1!) = 5$. On top of these, there is the single taste-explosion that comes from all five tastes being combined.

Adding these up, there are the five single flavours, the ten pairs, the ten triples, five quadruples, and the one flavour of everything. This totals $5 + 10 + 10 + 5 + 1 = 31$. In fact there is one more we could consider, the bland taste of nothing. Incorporating this, we get an answer of 32. This is mathematically interesting, as $32 = 2^5$, that is to say, $2 \times 2 \times 2 \times 2 \times 2$. When Sushruta considered the possible combinations of six tastes, he arrived at 64 (if we also add in the taste of nothing). This is of course 2^6. In general, the total number of combinations from a collection of n objects, is 2^n. This is an extremely useful fact, as this sequence grows very fast indeed, making direct counting almost impossible.

● The birthday problem

A famous question asks: how many people do there need to be in a room for there to be a reasonable chance that two will share the same birthday?

To make this something that can be attacked mathematically, we need to be more precise. Obviously, the only time we can be absolutely certain that two or more people will share a birthday is when there are 367 people in the room, meaning that at best 366 people can have birthdays on different days (including the leap day 29 February), leaving the final person to share her birthday with one of the others.

Let's suppose instead we want a probability of 50% or more that two people in the room will share a birthday. How many people need there be? To make life simple we will ignore leap years. We will also assume, slightly inaccurately, that all days are equally likely to feature as people's birthdays. A first guess at the answer might be 183 people, since this is just over 50% of 365. As it happens, the correct answer is much, much smaller.

It is convenient to turn the problem around, and ask what is the probability that all the people present will have different birthdays? Once we have found a number that gives an answer of less than 50%, we have found our solution.

Suppose there are two people present, Arnold and Betty. For each of them there are 365 possible birthdays, so the total number of possible pairings for the two is 365×365. How many of these represent different birthdays? Well, taking the people in order, Arnold can have his birthday on any day, giving 365 possibilities. But if the two are not to match, then Betty must avoid his birthday, giving her 364 possible birthdays, and a total of 365×364 pairs of non-matching birthdays. So the probability that the two have different birthdays is $\frac{365 \times 364}{365 \times 365}$, around 0.997.

This same reasoning extends to more people: the probability that ten people have all different birthdays, say, is given by $\frac{365 \times 364 \times 363 \times \ldots \times 356}{365^{10}}$, which is around 0.883. However, once we reach 23 people, the probability is $\frac{365 \times 364 \times 363 \times \ldots \times 343}{365^{23}}$, around 0.493. So with 23 people in the room, there is only a 49% chance that everyone will have different birthdays, giving a 51% chance that two people will share a birthday.

● Ramanujan's partitions

One of the greatest mathematical minds of all time belonged to Srinivasa Ramanujan. Born in rural India in 1887, Ramanujan excelled in mathematics from an early age, teaching himself from the few textbooks he could find. He quickly progressed to his own research, single-handedly rediscovering several great theorems in the process. Ramanujan struggled to find work as a mathematician in India, and in 1913 hopefully wrote to several British mathematicians. Foremost among them was G.H. Hardy, who concluded from the rather eccentric letter he received that he was dealing with 'a mathematician of the highest quality, a man of altogether exceptional originality and power'. In 1914, Ramanujan travelled to the University of Cambridge to begin a wonderful collaboration with Hardy.

One of the questions Ramanujan and Hardy considered was deceptively simple: how many ways are there to break up the number 4? We could write it as $1 + 1 + 1 + 1$, or $2 + 1 + 1$, or $2 + 2$, or $3 + 1$, or finally just as 4. This gives the number of possibilities as five, called the five *partitions* of 4. If we start with 2, on the other hand, it can only be broken down as $1 + 1$ or 2. The same question beginning with 5 produces seven partitions:

$$1 + 1 + 1 + 1 + 1 = 2 + 1 + 1 + 1 = 2 + 2 + 1 = 3 + 1 + 1 = 3 + 2 = 4 + 1 = 5$$

If we continue with this, a sequence develops: 1, 2, 3, 5, 7, 11, 15, 22, 30, 42, ..., which counts the number of partitions of 1, 2, 3, 4, 5, 6, 7, 8, 9, 10, ... respectively. What is the meaning behind this sequence? In particular, if we want to know the number of partitions of 100, how can we even begin to work it out?

This question has perplexed many great minds, and even the combined powers of Ramanujan and Hardy were not able to produce an exact formula. However, they did find a very good approximate answer. What is more it becomes ever more accurate, the larger the number we look at. They proved that the number of partitions of the number n is approximately:

$$\frac{1}{4n\sqrt{3}} e^{\pi \cdot \sqrt{\frac{2n}{3}}}$$

'I count a lot of things that there's no need to count... Just because that's the way I am. But I count all the things that need to be counted.'

Richard Brautigan, *The Hawkline Monster: A Gothic Western*

14 How to visit a hundred cities in one day

- Crossing the bridges of Königsberg
- Euler and the birth of graph theory
- Kwan and the Chinese postman problem
- The travelling salesman problem

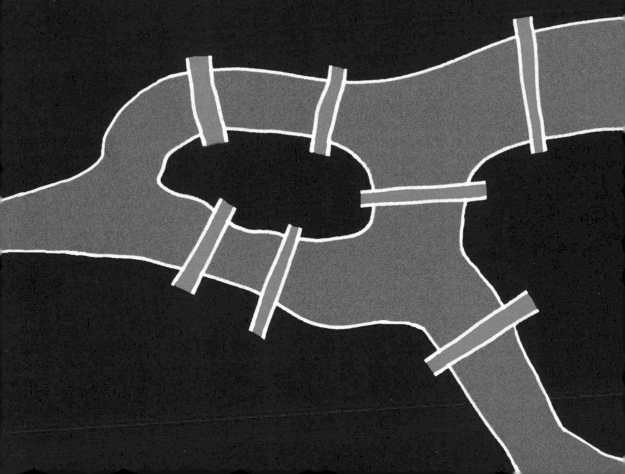

The Russian town of Kaliningrad has a rich history. Previously, it was known as Königsberg and was the capital of Prussia, part of the German Empire. Königsberg was an intellectual and cultural centre of Europe, and home to luminaries such as the philosopher Immanuel Kant, and the composer Richard Wagner. It was particularly famous for its mathematics, and was the birthplace of Christian Goldbach and David Hilbert among others. But it is not just for its illustrious residents that Königsberg was famous. In the early 18th century, the town itself became the focus of one of the most celebrated problems in the history of mathematics.

The problem was not deep or difficult, just a diverting puzzle that the townspeople used to play on a summer evening. Yet this little game opened the door to an entirely new type of mathematics, known as *graph theory*. This plays a central role in modern mathematics, and has numerous practical applications such as in the route-planning software employed by cars' satellite navigation systems.

Crossing the bridges of Königsberg

The city of Königsberg sits on the river Pregel. This river splits in the middle of the town, dividing the city into four parts, which are connected by seven bridges. The game that young Königsbergers would play was to try to walk around the city, choosing the route so that they cross each of these bridges exactly once. Try as they might, however, neither the brightest school students nor the wisest old heads could find a route that worked.

This little puzzle attracted the interest of the top mathematician of the age, Leonhard Euler. He answered the question with little difficulty, but more importantly made several observations that resulted in the birth of an entirely new branch of mathematics.

The first comment that Euler made was that the precise geography of the city was not relevant. In modern terms, all the information that mattered can be captured in a small diagram called a *graph*, where the four land-masses were represented by single dots and the bridges by edges between these dots. (In other areas of mathematics, the word *graph* also carries other meanings.)

This was a radically new approach to a geometric problem (which Euler credits to Gottfried Leibniz). The power of the graph comes from its

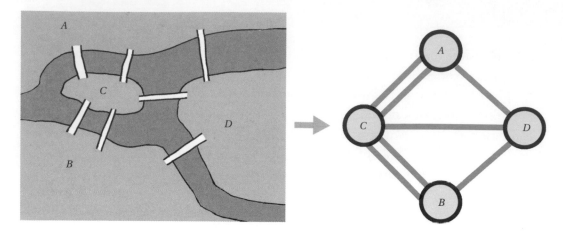

simplicity. All extraneous details are omitted: the curve of the river and the lengths of the bridges, and so on, leaving the core of the problem exposed. For this reason graphs are ubiquitous in mathematics today, and are particularly important in *topology*, where geometrical objects are considered the same if they can be stretched into each other (see *How to find all the holes in the Universe*).

The second observation Euler made quickly led him to the answer. The dots, or *vertices* as they are known, had different numbers of edges coming out of them. Now, when you travel around a graph, edges come in pairs. You arrive at a vertex through one edge, and leave it through another. The Königsbergers wanted to travel each edge exactly once, so the entrance and exit had better be different. What can go wrong is that we arrive at a vertex and get stuck, finding that there is no available exit that hasn't already been travelled. To avoid this, most vertices must have an even number of edges attached, so that for each entrance there is an available exit.

The number of edges at a vertex is called its *degree*. So Euler's criterion is that, for a suitable circuit to exist, the degree of every vertex must be an even number. If we don't mind the route beginning and ending at different points, then there may be two vertices that have odd degree, while all the rest are even. In the Königsberg graph all four vertices have odd degree. So no such path can exist.

Euler's answer to the bridges of Königsberg is enough to solve another popular type of puzzle: whether a certain figure can be drawn without taking your pen off the page or repeating any lines (or employing any tricks such as going round the back of the page). If we consider the picture as a graph, then Euler points us to the critical question: do all the vertices except two have an even degree? If so, then there is a route around the graph that travels each edge exactly once: this is precisely what we want. If not, then, as for the problem of the bridges, there is no solution.

● The Chinese postman

Two of the centrepieces of modern graph theory are the Chinese postman problem and the travelling salesman problem. These two seem similar at first, but theory reveals them to be entirely different at a deeper level.

Suppose a postman needs to deliver mail to every street within his region of town. When he plans his route, he wants to minimize the total distance he needs to walk every day. It is worth his while spending some time considering this, since the difference between an optimal route and an inefficient one could be considerable. The crux of the problem can again be analyzed using graph theory.

In this case the edges of the graph will represent the streets of the area. The vertices represent the junctions where different streets meet. The first question is exactly the same as for the bridges of Königsberg: is there a route around the graph that travels each edge exactly once? The answer, as before, will depend on the degree of the vertices. If every one has even degree, then there is such a route, and if not, then there is not.

That may be an interesting fact for mathematicians to mull over, but if the answer is no, as it is quite likely to be, it doesn't much help the postman. After all, he still has his bag of letters to deliver. In this case he will have no option but to travel some of the roads more than once. But which ones should he choose? Another factor comes into play here: some roads are longer than others. To capture this, the graph acquires an extra ingredient. Now every edge is assigned a number called its *weight*. This may represent the length of the road, or the time it takes to travel it. The two are not always the same: imagine driving 10 miles down a farm track instead of 11 along a motorway. In other applications weight can represent other quantities, such as the bandwidth or financial cost. In any case, the postman's problem amounts to finding a route around the graph, travelling each edge at least once, keeping the total weight to a minimum.

This is the *Chinese postman problem*, so-called for no reason other than the fact that the first person to study it was a Chinese mathematician Mei-ko Kwan in 1962. Kwan found a way to solve the problem: he provided a simple set of instructions (or an *algorithm*, see *How to bring down the internet*) that

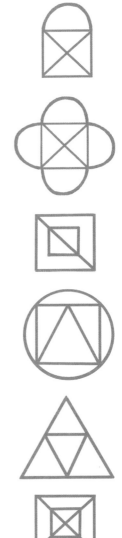

Can you draw these without taking your pen off the page?

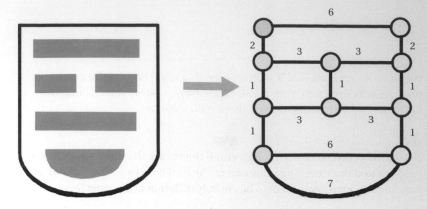

will always produce an optimal solution to the Chinese postman problem.

Kwan's solution amounts to pinpointing the vertices that have odd degree, and expanding the graph by connecting them in pairs. These new edges represent routes along the original edges, and had to be chosen of minimal weight. With this done, the vertices in the newly expanded graph would all have even degree, so a route around must exist.

With the aid of a computer, Kwan's algorithm quickly homes in on an optimal solution for the postman's morning route. The same cannot be said of this problem's bigger brother, the travelling salesman problem.

The travelling salesman problem

Suppose a travelling salesman needs to visit 10 different cities. He knows the distances between every pair of cities, and wants to plan his route to minimize the total distance travelled. This type of question is tailor made for graph-theoretical methods. As before, the scenario is modelled as a graph: the ten cities are represented by the vertices of the graph, with the edges being the roads between them, and the weights of the edges representing their length. What is required is a route around the graph that visits every vertex once and has the minimal total weight.

The resulting problem may seem similar to the Chinese postman problem. However, there is a small distinction, namely that the salesman no longer needs to travel along every edge of the graph. He only needs to visit every vertex. As it turns out, this subtle change makes a massive difference to the problem, increasing its complexity by a huge amount.

While Kwan had been able to find a quick and efficient algorithm for solving the Chinese postman's problem, there can be none such for the travelling salesman problem.

Of course there is a way to find the optimal route: you can list every possible route the salesman could take, calculate the weight of each, and pick the lowest. This will certainly work, eventually. The problem is that the number of possible routes is very large. For an 11-city graph, there may be over a million routes to consider, and this rapidly increases. A 61-city graph may potentially contain more routes than the number of atoms in the universe.

The obvious question then is whether there is a better algorithm to solve the travelling salesman problem: a clever piece of trickery such as Kwan found in the Chinese postman's case? In this case, however, the answer is very likely to be no. In the jargon of modern complexity theory the problem is *NP-complete*, which essentially means that it is irreducibly complex. There can be no way to solve it without having to perform a very large number of intermediate calculations. There is a glimmer of hope for a fast solution to the travelling salesman problem, but it is faint indeed: if someone could prove that $P = NP$, that would mean that there is after all a quick way to solve it (see *How to bring down the internet*).

The travelling salesman keeps travelling

Many questions in science and technology involve optimizing some quantity, whether it is the speed of a computer processor, the cost of manufacturing a television, or the amount of traffic crossing a set of traffic lights. The beauty of graph theory is that it does not matter what physical quantity is involved; they can all be studied in exactly the same manner, as weights on a graph. Many such optimization problems can then be encoded as Chinese postman or travelling salesman problems, so these two innocent-sounding problems of are of huge practical importance in the world today, from telecommunications to genome sequencing.

Although there is no way to solve the travelling salesman problem exactly, several algorithms have been found that can solve it approximately. One such algorithm is to tell the salesman to travel to the nearest city at each step, avoiding any that he has already visited. This is a surprisingly good method, but more sophisticated algorithms can do better.

In 1991, Gerhard Reinelt put together a library of travelling salesman problems from a range of sciences and industries, and challenged the mathematical community to solve them exactly. The largest solved so far is an 85,900-city problem, completed in 2006 by David Applegate and colleagues. The computation required a total of 136 years of computer processing.

15 How to arrange the perfect dinner party

- The dinner party problem
- Order, disorder and Ramsey's theorem
- Convex and non-convex shapes
- The happy ending problem

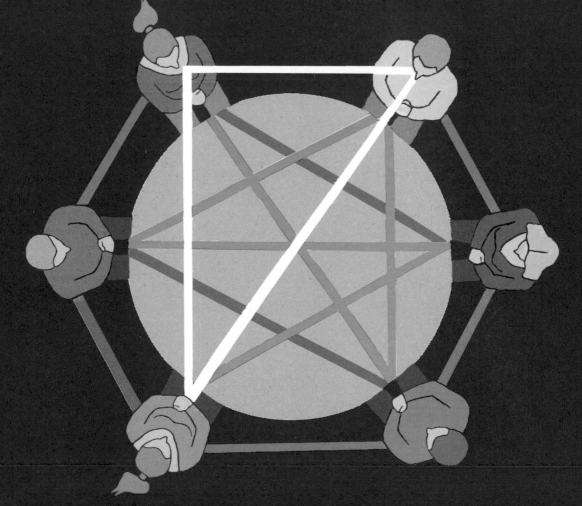

There are innumerable theories about what makes a good dinner party: gourmet food, vintage wine, a string quartet playing in the hallway. Well, there are other books you can consult for advice on these subjects. But mathematics does have something to contribute to the discussion, in terms of choosing the right mix of company. Remarkably, the question that this provokes is one of the most formidably difficult in the subject.

The critical question, from a mathematical perspective, is how many of the guests know each other. If everyone present knows each other, then a pleasantly relaxed evening should ensue. On the other hand, a party of total strangers can also be successful, where people are simply thrown together and left to get to know one another. In between these two extremes, problems can occur. If everyone knows each other except for one person, then he may feel excluded from the fun. If one person is known to everyone, but none of the others know each other, then she may become the centre of attention.

One of the tasks facing the party organizer, therefore, is to gauge the balance of friends and strangers among the guests. Of course, in social terms, this is something of a crude simplification. Nevertheless, the mathematics at the core of this problem is of astonishing complexity. Indeed the right way to formulate a dinner party was a major mathematical question of the 20th century, and is still unresolved today in many cases.

The dinner party problem

To bring mathematics into the picture, we will make several simplifying assumptions: people are either friends or strangers (there are no shades of grey where people know each other slightly). We will also make the utopian assumption that all people who know each other are friends. There are no enemies in this universe, only strangers. Finally, friendship is always two-way: if Alex knows Belinda, then Belinda must also know Alex.

Now here is a question: how many guests need to be invited to ensure that there are either three mutual strangers or three mutual friends present? Replacing 'three' with 'two', the question has a simple answer: only two guests are needed (they either know each other or they do not). But if we invite three guests, it might be that Alex knows Belinda, and Belinda knows Charlie, but Charlie does not know Alex. In this case, we have neither three mutual friends, nor three mutual strangers. So, three guests are not enough.

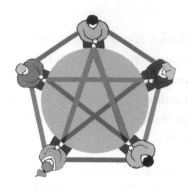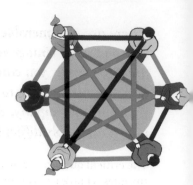

To investigate further, it is helpful to draw diagrams of the dinner parties. If two guests know each other, then they are connected with a blue edge. If they do not, then they are connected with a black edge. (So every pair must be connected by an edge of one colour or other.)

The question now becomes: how many are needed in total, to ensure that, however you join them up, there must either be a black triangle or a blue one? Four guests are still inadequate: it is possible to construct a party of four that includes neither a blue triangle nor a black one. The same is true for parties of five. But in any party of six, no matter what the arrangement of edges, a triangle of one or other colour must occur. So the answer to the original dinner party problem is six.

How complicated can a dinner party be?

There is a natural way to extend the dinner party problem: we can increase the required number of mutual friends or strangers to four. A diagram where four people are all connected is called a *4-clique*. So the question is: how big a diagram is needed to ensure that it contains either a black or a blue 4-clique. This hugely increases the complexity of the problem, as the number of different possible diagrams that need to be checked increases quickly. There are many billions of different possible diagrams of size 10 for example, and even the fastest supercomputer will quickly become overloaded.

In 1955, Greenwood and Gleason found that the threshold in this case is 18 people. Any dinner party of 18 or more people must contain either four mutual friends, or four mutual strangers. But if we raise the required number to 5, the number of possible diagrams becomes totally unmanageable. In fact, at time of writing, the answer to this question is still unknown. In the last twenty years, work by Exoo, McKay and Radziszowski has pinned it down to somewhere between 43 and 49 guests, but the exact value remains to be discovered.

How to arrange the perfect dinner party

We can extend the question even further, by dropping the symmetry. For example, we could equally well ask how many guests are needed to guarantee that the party must include either five mutual strangers or three common acquaintances. Greenwood and Gleason found that the answer is 14. What about seven mutual strangers or eight common acquaintances? The exact value is unknown, but is somewhere between 216 and 1,031.

Frank Ramsey: every dinner party is possible

All of these dinner party questions were simultaneously addressed by a seminal result, called *Ramsey's theorem*, after its discoverer, Frank Ramsey. Ramsey was a prodigious talent who made significant contributions to not only mathematics, but also philosophy and economics. A tall man, with an infectious laugh, he quickly rose to a lectureship in mathematics at the University of Cambridge, where he was renowned for the clarity and enthusiasm of his talks. In 1930, the same year that his most famous theorem was published, Ramsey fell ill with jaundice, and died at the tragically young age of 26.

Among other important consequences, Ramsey's theorem guarantees that each dinner party problem must have a solution. This may not sound a spectacular insight, but in mathematics nothing is considered certain until it is proved. Before Ramsey, it was conceivable that there could be parties of arbitrarily large size that fail to have some particular configuration of friends or strangers. Thanks to Ramsey's theorem, we know that there is some threshold number of guests that will guarantee either seven mutual strangers or five common acquaintances (recently pinned down to between 80 and 143) and another threshold for eight strangers or twenty friends (somewhere between 1,094 and 657,799 guests).

Ramsey's theorem has applications throughout mathematics, and it was especially important in this case, given the incredible difficulty of pinning down the required sizes of the dinner parties.

'The best number for a dinner-party is two: myself and a damn good head waiter.'

NUBAR GULBENKIAN

Ramsey theory: order among disorder

The subject of Ramsey theory was born with the dinner party problem, but has subsequently grown to encompass many other deep and difficult questions. The general phenomenon considered by Ramsey theory is the emergence of order from disorder. This is exemplified in the dinner party problem. The most ordered and structured types of diagram are the ones with just one colour, the single coloured cliques.

The dinner party problem manages to extract this type of order from disorder. No matter how disordered and unstructured the diagram of an 18-person party is, there is guaranteed to be a part of it that is highly ordered, namely either a blue or black 4-clique. This type of result is hugely important as it helps tame even the wildest mathematical structures, guaranteeing that they have substructures that are well-behaved and rigidly organized. This is the power of Ramsey theory. Another important example is the happy ending problem.

In general position; 4 points form a convex quadrilateral

Dot-to-dot puzzles are more difficult than you think

Suppose we start with a blank piece of paper, and idly start drawing dots on it at random, without worrying at all about where they go. The first way that order can emerge from this situation is if some dots lie in a straight line. Of course, any two dots automatically lie on a straight line. For three or more dots to lie exactly on a straight line is statistically highly unlikely, so we assume this does not happen. A collection of dots, where no three lie on a straight line, is said to be 'in general position'. These are the collections we are interested in.

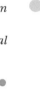

Not in general position

As soon as we have four dots we can join them together to get a four-sided shape, a *quadrilateral*. Quadrilaterals come in two essentially different forms. The nicest types, from a geometric point of view, are the *convex* quadrilaterals. A shape is *convex* if whenever you pick two points and join them with a straight line, that line lies entirely inside the shape. An arrowhead, for example, is non-convex. The emergence of convex quadrilaterals represents order in these situations.

A convex quadrilateral and a non-convex quadrilateral

It is not true that any four points can be connected to form a convex quadrilateral. However, in 1932, Esther Klein made the important observation that whenever you draw five dots on a piece of paper (assuming as usual that no three lie on a straight line), then there must always be four among them that can be joined together to form a convex quadrilateral. Klein was able to prove that this should always be the case, no matter how baroque the arrangement of five points.

A happy ending

To a mathematician's mind, Klein's dot-to-dot theorem prompts an obvious next question: how many points are required to guarantee the existence of a convex pentagon? What about a convex hexagon, or in general a convex *n*-gon? This is more difficult. This question was dubbed the *happy ending*

> *'Confusion heard his voice, and wild uproar*
> *Stood ruled, stood vast infinitude confined;*
> *Till at his second bidding darkness fled,*
> *Light shone, and order from disorder sprung.'*

JOHN MILTON, *Paradise Lost*

problem by Paul Erdős because the two researchers who began investigating it, Esther Klein and George Szekeres, ended up not only discovering some wonderful mathematics, but also getting married.

As with the dinner party problem, the exact numbers are exceptionally difficult to pin down. With a little investigation, Endre Makai was able to prove that the number of points needed to guarantee a convex pentagon is nine. Eight is not enough, as the illustration shows. Then it took until 2006 for Szekeres and Lindsay Peters finally to show that the number of points required to guarantee the existence of a convex hexagon is 17. Beyond this, the numbers remain mysterious. However, Szekeres and Paul Erdős did manage to use Ramsey theory to prove the critical result that these questions must always have an answer. There will be some number, so that any arrangements of points (no three of which lie on a straight line) must contain a convex heptagon.

These Ramsey-type results, including the problem of the dinner parties and the dots on the page, are profoundly important. Their moral is that, even amidst the most total confusion and disorder, we can always find islands of stability, organization and order. It is a momentous and inspiring idea, not to say an exceptionally useful one within mathematics and beyond, to computer science and artificial intelligence research. More surprising is the fact that we flirt with this very notion every time we plan a dinner party, or join together a few dots on a page.

8 dots containing no convex pentagon

16 How to paint the world in four colours

- The map-colouring problem
- Planar graphs and gas, water and electricity
- Appel and Haken's gigantic proof
- The birth of computational mathematics

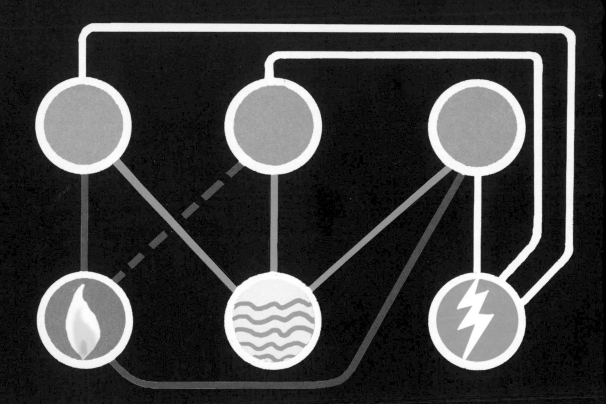

How can you sell painting equipment to people with little artistic ability? In the 1950s, a paint manufacturer had an idea. The product was called painting by numbers. The kit comprised a selection of paints, each of which was numbered (1 for crimson and 17 for turquoise, for example). Along with the paints came brushes and a canvas, which had already been divided up into regions. Each of these regions was numbered. To complete the painting all you had to do was colour each region with the paint of the corresponding number. The point of painting by numbers is that it is easy to do. It is surprising, then, that underlying it is a piece of mathematics so deep and difficult that it challenged the very nature of the subject.

For mathematicians, the interesting question is: what is the minimum number of different coloured paints that are enough to fill in any possible canvas? Obviously, the answer is one, we could just paint every region pink, for example. It becomes more interesting when we impose the following rule: two regions that border each other may not be painted the same colour. Regions that only touch at a single point are permitted to be the same colour, however. So for a chessboard pattern, two colours are enough. But it is easy to come up with designs that require three or four colours.

Guthrie's map-making experiments

Francis Guthrie was a South African who travelled to London to study mathematics and law. Guthrie did not have painting by numbers in mind (it had not been invented), but rather the colouring of a map. How many different colours are needed to paint a map so that no two countries that share a border are the same colour? Guthrie began experimenting with map-colouring, and in 1852 he came to the conclusion that the maximum number of colours that was ever needed was four. However elaborate a map he drew, he was unable to make one where five colours were needed. (We assume all countries come in one piece, rather than disconnected parts as the USA does, with Alaska forming a separate block.)

It is one thing to experiment and make such a claim. It is another matter to prove it. Only with a watertight mathematical proof could Guthrie be sure that he had not missed some particularly convoluted map requiring five different colours to paint. Unable to find such a proof, Guthrie consulted his brother Frederick, also a mathematician. The problem soon found its way to the desks of the top mathematicians in Britain, including Augustus de Morgan, Arthur Cayley and William Rowan Hamilton. When these luminaries were also unable to produce a proof, the four colour problem had grown from an intriguing puzzle into a serious mathematical problem. In retrospect, we can see today that even the greatest thinkers of the age never stood a chance.

Gas, water and electricity

One of the oldest mathematical puzzles of all concerns three houses that each need to be connected to three utilities: gas, water and electricity. Is it possible to arrange the pipelines so that they never cross over?

To turn this problem into a mathematical one, we interpret the three houses and three utilities as dots on a page. The pipelines are then lines connecting the dots. A little experimentation shows that it is easy to connect two houses to the three utilities (or vice versa). But, however you try to connect the final pair, its pipeline will always cross another.

An arrangement of dots connected to each other with edges is known to mathematicians as a *graph*. These objects appear throughout mathematics (see *How to visit a hundred cities in one day* and *How to arrange the perfect dinner party*). One intriguing question is: when is it possible to represent a graph on a piece of flat paper without the edges crossing? The three-utilities graph is the first example of a graph that cannot be represented in this way.

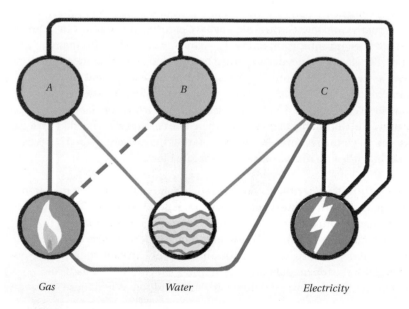

Gas   Water   Electricity

Another type of graph is formed by drawing a collection of dots on the page, and then trying to connect each of them to every other. Such a graph is called *complete*. If you try this with three points, an ordinary triangle does the job. With four points, it is possible too. But when we get to five, however hard you try, you will always end up with edges that cross.

Three point complete graph

This gives us two different graphs that cannot be drawn on a flat piece of paper without crossing each other: the three utilities graph, and the complete five-point graph. In technical terms they are *non-planar*. Of course, if any bigger graph has either of these two as a subgraph, then it cannot be planar either. More surprising is that this is essentially the *only* barrier to being planar. A seminal result of 1930 called Karatowski's theorem shows that if any graph is not planar (that is, it can never be drawn on a flat piece of paper without its edges crossing), then it must somehow encode either the complete five-point graph or the three utilities graph.

Four point complete graph

● Proofs, disproofs and reproofs

What is the relation between these dot–line graphs and the four colour problem? Well, start with a map and place a dot at the centre of every country. Now, if two countries share a border, connect the two dots. With this done, we can get rid of the countries altogether, because the essential information is encoded in the graph.

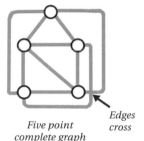

Five point complete graph

Edges cross

Having done this, the study of planar graphs contributes useful data. Since we know that a five-point complete graph can never be drawn on a piece of paper, it follows that no map can exist where five countries all touch each other. Doesn't this mean that the four colour theorem must be true? Not so fast! It is certainly the case that any configuration of five countries can be coloured with four colours, when taken in isolation. The difficulty lies in what happens outside this small configuration. So, to solve the problem, it is necessary to analyze the possible arrangements beyond a region of five countries, in particular the way that the countries can affect each other indirectly for example through chains of alternating colours.

Creating a graph from a map

This was the approach taken by a 19th-century lawyer, Alfred Kempe. Kempe had studied mathematics as a young man, and devoted much time to the subject throughout his life. In 1879, after some investigations Kempe announced that he had a proof: four colours are indeed enough. Kempe's solution turned him into something of a mathematical celebrity, featuring in *The Nation* magazine

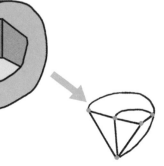

and helping him gain election to the Royal Society. Sadly for him, however, his argument was flawed, as was pointed out by Percy Heawood in 1890.

● Heawood: five colours are enough

Athough Kempe's proof contained a crucial gap, Heawood was able to build on his ideas to prove that five colours were certainly enough. No map would ever be found that required six. This proof of the five colour theorem went as follows: Heawood reasoned that if there is a counterexample (that is to say, a non-5-colourable graph) then there must be one of minimal possible size. So if he started with this, and then removed one point, the remaining graph must definitely be 5-colourable. He took this 5-colouring and needed to find a way to reinsert the final point. Building on Kempe's work, Heawood was able to show that it could be done, meaning that every graph is 5-colourable.

● Appel and Haken, and the power of the computer

That is how the problem remained for most of the next century. No-one was able to improve on Heawood's theorem, and reduce the number to four. Of course, there was no shortage of people willing to try, attracted by the misleading simplicity of the problem. Nor was anyone able to construct a new form of map that required five colours.

It was not until 1976 that the long-awaited proof finally arrived, courtesy of Kenneth Appel and Wolfgang Haken at the University of Illinois. The guesses and conjectures had all been right. As they triumphantly announced on the University of Illinois website, 'Four colours suffice'.

If the final result was as expected, no-one could have guessed what the eventual proof would look like. In principle, their approach was the same as that of Kempe and Heawood, but turbo-charged to an extraordinary degree.

The argument was subtle: first they found a list of different configurations of countries, such that any planar graph must contain one of them. Then they took each configuration in turn and performed the following piece of analysis. They assumed that the whole graph apart from the given configuration can be 4-coloured, then showed that it must also be possible to recolour the entire graph using four colours, including their configuration.

This would be enough to prove the theorem. If there is any graph that cannot be 4-coloured then there must be a smallest possible one. What is more, this graph must contain one of their special configurations. If we remove that portion, then what remains must be 4-colourable (because the starting graph was assumed to have been a smallest possible non-4-colourable graph). But then, according to their proof, it must be possible to recolour the whole graph using four colours, including the configuration. So the result is that the graph we started with is 4-colourable after all.

If the idea was subtle, actually performing the necessary analysis was another matter altogether. Whereas Kempe and Heawood had needed to consider only five different configurations, Appel and Haken needed to check a total of 1,936. Worse than this, verifying the 4-colourability of some of these configurations entailed recolouring hundreds of thousands of broader arrangements. The result was the longest mathematical proof there had ever been. It is no surprise that Appel and Haken needed to exploit the full power of the modern computer, in addition to their mathematical talents. In all, their proof required over 1,000 hours of processing time.

Computers versus mathematicians

Since Appel and Haken's breakthrough, the proof of the four colour theorem has been simplified somewhat, though still not to anything digestible by a human alone. It remains disconcerting that such a gargantuan proof should be required to prove something that an intelligent adult would guess after a few minutes experimentation.

Indeed, Appel and Haken's work was a challenge to the very nature of mathematics. Since the first one was written down, a mathematical proof has been the quintessential expression of human understanding. In proving a fact, we are showing that we have fully mentally mastered it. A proof is a way to make this understanding explicit, and thereby certain, and communicable to others. Implicit in Haken and Appel's work was a demand that we should cede some of the responsibility for proving a theorem to the computer. Many mathematicians found this difficult to stomach, but the fact is that until someone comes up with a more concise proof, we simply have no choice in the matter: we have to trust our computers. The proof of the four colour theorem was the first herald of an extraordinary fact: that the future of mathematics is certain to be inextricably bound up with the development of artificial intelligence.

'I visualize a time when we will be to robots what dogs are to humans. And I am rooting for the machines.'

Claude Shannon

17 How to be alive and dead at the same time

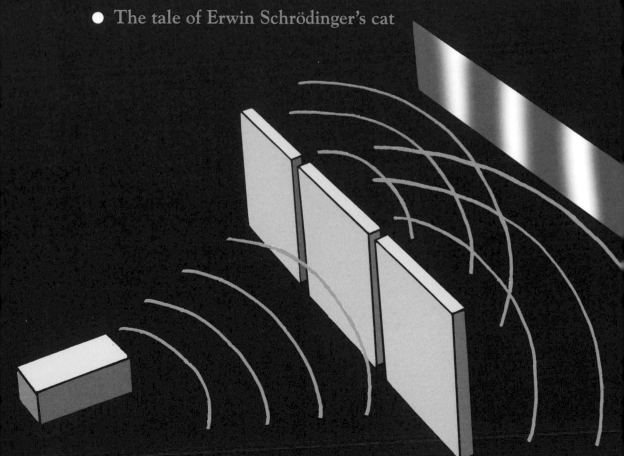

One of the most famous of all scientific experiments is the double-slit experiment, first performed in 1801, by Thomas Young. He was attempting to answer one of the burning scientific issues of the age: does light come in waves or particles?

Young shone a light at a screen. In between the light and the screen was a wall, with two slits in it. What would he expect to see on the screen? The answer would depend on the nature of light. If light consists of particles, then Young should just see a concentration of light behind each slit, and a darker region between them, a fairly simple scenario. But if light comes in waves, he should see something else, because waves interfere with each other. This can be seen clearly with water waves. Imagine a child at one end of an otherwise empty and still swimming pool, holding onto the side and practising her kicking. Through the movements of her legs, a smooth wave is set up across the pool, the surface rising and falling with perfect regularity. However if another child starts practising his kicking at another point along the edge of the pool, the motion of the water becomes more complicated. The two waves reinforce each other at some points leading to high peaks and deep troughs, and cancel each other out elsewhere producing flat water. The resulting picture is called an *interference pattern*.

When Young performed the double-slit experiment, he saw two main concentrations of light behind each slit, but between them a pattern of stripes, alternating between dark patches and bright ones. This was an unmistakable fingerprint of waves interfering. The beams of light passing through the two slits were reinforcing each other at the light patches, and cancelling each other out at the dark.

What Young saw and what he might have seen

The birth of quantum theory

For around 100 years, the wave–particle debate about the nature of light faded, Young having apparently settled it in favour of waves. Actually, the story had only just started. In the early 20th century, evidence of a very different kind was uncovered by Max Planck and Albert Einstein. When light is shone at a piece of metal, small particles called electrons are emitted from it, as if knocked out by the incoming light. The puzzling thing was that if light is a wave, we would expect the speed of the expelled electrons to depend on the energy of the light, that is to say its intensity. The more intense the light is, we would expect correspondingly faster electrons coming out, just as a balloon will fly faster in a strong wind than in a light breeze.

But this was not what actually happened. Instead a greater intensity of light produced more electrons, but each one had exactly the same speed. This puzzling observation was dubbed the *photoelectric effect*. It was Albert Einstein who resolved the paradox, by reviving the old particle-theory of light. He suggested that the electrons were being knocked out by an individual packet or 'quantum' of light, each carrying the same amount of energy. These packets were later dubbed *photons*.

The double-slit experiment (remixed)

Einstein's quantum theoretic explanation for the photoelectric effect was a triumph. Indeed it was this, not his theory of relativity, that earned him the Nobel Prize in 1921. It raised an important question, however: where did the new theory of photons leave Young's double-slit experiment?

Using more sophisticated apparatus than Young had available, it was possible to repeat his experiment, this time firing just one particle at a time at the screen, and observing where it ended up. The result of this experiment was one of the most shocking in the history of science. As more particles were fired at the screen, a picture gradually emerged. Again, it clearly bore the signature of an interference pattern. But if the particles were coming through one at a time, what could be interfering? Was it not the case that each particle must pass through either one slit or the other, with no interference possible? Apparently, it was not.

Even more surprisingly, when one of the slits was closed, the interference patterns disappeared. The dark patches became brighter again. Somehow it seemed that opening up a second slit was actively preventing a particle from arriving at a previously accessible point.

Photons interfering

Does God play dice?

How can we account for the strange observations of the double-slit experiment? It seems that light comes neither as waves nor as particles, but as something else, which in different scenarios can exhibit the properties of both.

This *wave–particle duality* is extremely difficult to understand from a psychological standpoint. However, there is a toolbox that scientists can use to get to grips with it: mathematics. The approach that physicists adopt is simply to use the mathematics to make predictions, and test those predictions against experiment. Sometimes science can progress without humans having to worry about what it all means.

Luckily, the mathematics of quantum mechanics is not as confusing as the physical reality, at least not for the experts. For the layperson it still provides a challenge. The fundamental mathematical device is that of a *wave function*. This is the mathematical description of wave–particle duality. Usually denoted by the Greek letter psi, ψ, a wave function assigns a number to every region of accessible space, which encodes the probability of finding the particle there. So if ψ has a value of 0.5 at one place and 0.001 at another, then the particle is much more likely to be found at the first than the second.

So far, this could be described by plain probability theory, which had been known for centuries. However, wave functions do something that ordinary probability does not: they interfere, as waves do. To allow this to happen, the numbers assigned by ψ need to be complex numbers (see *How to solve every equation there has ever been*).

This is the modern resolution to the ancient riddle of the nature of light: it is described by a complex wave function ψ. This had the unexpected effect of introducing uncertainty at the most fundamental level of reality. Einstein for one did not like this, insisting that 'God does not play dice'.

What's your wavelength?

It is not just light that has this unfamiliar nature, something between waves and particles. In the same way that Einstein and Planck had hypothesized that light might have a particulate side to it, Louis de Broglie suggested that ordinary matter, traditionally viewed as being made of particles, might have a wavy side. De Broglie came up with a simple equation to translate between the language of particles and waves. If a particle has mass m, and is travelling at speed v, then the wavelength of its associated wave is given by $\frac{h}{m \times v}$ where h is the so-called *Planck constant*, with a value of around 6.6×10^{-34} (that is, 0.000 . . . 00066, where there are 34 zeroes in total). It is because this value is so incredibly small that the wavy nature of matter is invisible to us. Nevertheless it has been verified experimentally, not least by repeating the double-slit experiment using particles of matter, and seeing exactly the same interference patterns.

A quantum universe

A wave function ψ can describe any particle of matter or light (a photon). More complex wave functions can also describe larger quantum systems consisting of many interacting particles. In principle there is no reason why this should not be true for the whole universe.

Neither traditional waves nor particles are static; the whole point is that they move and vary over time, and there are areas of classical physics dedicated to understanding these types of change. Similarly, a wave function ψ needs to evolve over time. This is governed by the fundamental equation of quantum mechanics: $i\hbar \frac{\partial \psi}{\partial t} = H\psi$ formulated by Erwin Schrödinger in 1926, which earned him a Nobel Prize. With this discovery, the mathematics of quantum mechanics could really get underway.

To understand how a quantum system stands at a certain moment, mathematicians need to find its wave function at that moment. The way to do this is to solve the Schrödinger equation. It is comforting that the philosophical peculiarities of quantum mechanics can now be left behind, and instead we just need to solve an equation. This, of course, is something that mathematicians are adept at.

Quantum hide-and-seek

Quantum physicists generally view particles as being spread out over a broad region, the exact spread being described by the wave function ψ. Among other things, this encodes the probability of finding the particle at any particular place. Suppose that the probability of finding the particle at some location is 50%. I have a look, and the particle is indeed there.

What is strange about this is that the probability is now no longer 50%, but 100%. By looking at the system I have fundamentally changed the wave function. Instead of being spread out around the room, it has become concentrated in one place.

This process is called the *collapse* of the wave function. The disconcerting fact is that it seems only to happen when someone or something observes the system. This uncomfortable conclusion is known as the *measurement paradox*. This is what happened during the double-slit experiment: the particle was released from the source, and became spread out, with two branches passing through the two slits, and interfering on the other side. Then someone had a look at the screen, and the wave function collapsed, revealing the particle somewhere there.

'If anybody says he can think about quantum physics without getting giddy, that only shows he has not understood the first thing about it.'

NIELS BOHR

How to be alive and dead at the same time

The cat of the living dead

The predictions of quantum theory are certainly unsettling. At least we can comfort ourselves that they are confined to the realm of small particles, and unlikely ever to affect our daily lives. Or perhaps we can't.

Despite being one of the principal characters in the development of quantum theory, Erwin Schrödinger was unsettled by the implications of his work. In particular he devised a thought experiment that showed how quantum phenomena can apparently lift from the world of particles to larger objects, and even involve living creatures.

Imagine a cat trapped in a box for an hour (this is purely a thought experiment, no real cats were harmed!). The walls of the box are very thick, so it is impossible for anyone outside to hear or otherwise tell what is happening inside. Along with the cat is what Schrödinger called a 'diabolical device', which is indeed worthy of a modern horror film. It consist of a small amount of radioactive material connected to a Geiger-counter that will detect if even a single atom of the material decays. The Geiger-counter in turn is connected to a hammer, which is positioned over a glass phial of poisonous gas. If it the counter is activated, the hammer is dropped, the phial smashed, and the gas released, killing the cat. The amount of radioactive material is carefully weighed so that there is a 50% chance that this will happen over the course of the hour.

The individual atoms of the radioactive material are quantum entities, and so, over the course of the hour, they will become spread out into a decayed/non-decayed state. The implication of this is that the Geiger counter will be in a quantum activated/non-activated state, the hammer in a quantum dropped/non-dropped state, the phial in a quantum intact/smashed state, the gas in a quantum released/unreleased state, and therefore the cat in a quantum living/dead state. This endures until someone opens the box to observe what is inside, and collapses the wave function of the entire system. Only then does the unfortunate cat settle down into a traditional state, either emerging from the box alive and well, or stone dead.

Schrödinger thought this conclusion 'ridiculous'. Along with Albert Einstein, he was convinced that quantum mechanics must be an incomplete theory. Some new factor would be discovered to clarify the limits of quantum phenomena. However, in the 75 years since Schrödinger first considered his unlucky cat, no-one has found a flaw in his reasoning, and the correct way to resolve this strange situation is not yet known.

'Quantum physics thus reveals a basic oneness of the universe.'

ERWIN SCHRÖDINGER

18 How to draw an impossible triangle

- The axiomatic approach of Euclid's *Elements*
- The parallel postulate
- The discovery of non-Euclidean geometry
- Hyperbolic and spherical geometry

Around 300 BC, Euclid of Alexandria wrote one of the most important books of all time, The Elements, which truly contains the wisdom of the ancient world. When it was translated into Arabic around AD 800, The Elements transformed Islamic mathematics. From there it returned to Europe, where it was translated into Latin in 1120. There it came to occupy a central place in the scientific renaissance, acting as an inspiration to the great scientists of the era, including Galileo and Newton. Over one thousand editions of The Elements have been produced, more than any other book except the Bible. For over 2,000 years, it remained a standard textbook, in use around the world.

The Elements was not all Euclid's original work. He combined his own insights with those of others to produce the definitive guide to mathematics. It contains important insights into prime numbers among other things, but the majority of the text is on the subject of geometry. Even today, the theorems about circles and triangles that are taught in classrooms around the world are essentially unchanged from Euclid's versions.

To begin at the beginning

The Elements is important not only for the theorems it contains. Euclid's organizing scheme for the book came to have an even more profound impact on mathematics. Euclid realized that mathematical facts and theorems are not heaven-sent, but need to be deduced from simpler results. So he organized the results in the book to develop from the simple to the complex. First he proved the most basic facts about lines and angles, building up through increasingly subtle facts about circles and triangles, and culminating in the classification of the *Platonic solids* (see *How to feel at home in five dimensions*).

Even the simplest results cannot be deduced from thin air, however. Right at the beginning, Euclid wrote down his starting assumptions, or *axioms*, on which all the results contained in the rest of the book ultimately relied. This philosophy, of building mathematics from the ground up, and explicitly stating the underlying assumptions, has been of unmeasurable imporance in the history of the subject.

Euclid sets the foundations

Euclid's principal purpose was to study *plane geometry*: configurations of points, lines, and circles on a flat sheet of paper. To this end, he began with five fundamental assumptions, known as *Euclid's postulates*:

1. Any two points may be connected with a stretch of straight line.

2. Any stretch of straight line can be extended indefinitely in either direction.

3. Given any point, you can draw a circle with this point as its centre. The circle can have a radius of any length.

4. Any two right angles are equal.

5. Given a straight line, and a point not on it, there is exactly one line passing through the point that is parallel to the line.

In fact, the fifth postulate, known as the *parallel postulate*, is not given here in the same terms that Euclid wrote it. Euclid's formulation was more cumbersome, although logically equivalent. His struggle to find the best formulation was the first stage in this law's extraordinary and controversial history. Certainly, Euclid preferred to work without it, and delayed relying on it for as long as possible. But when he came to prove a theorem well known to today's school students, he found that postulates 1 to 4 were not enough. He also needed to assume the parallel postulate.

That result is known as the *corresponding angles theorem*. It says that if a pair of parallel lines is cut by a third line, then the angles formed at corresponding positions on the two parallel lines will always be equal.

● Two thousand years of parallel lines

It is difficult to see why Euclid's parallel postulate was a source of heated debate for thousands of years. In one sense it is an entirely straightforward proposition. A little experimentation with a pen and paper can quickly convince anyone of its simple truth. To understand the controversy, we have to remember the purpose of Euclid's axiomatic approach.

The postulates were intended to form the underlying platform to support all the subsequent theorems. The question was: were the axioms really as efficient as possible? Were these five assumptions the minimum possible, or was there any hidden redundancy among them? It was the fifth postulate that fell under the spotlight.

Many people came to believe that the parallel postulate was not a necessary assumption, but that it was an automatic consequence of the other four. From Ptolemy in second-century Alexandria to Omar Khayyám in 11th-century Persia and Adrien-Marie Legendre in 19th-century France, this question became a preoccupation for many eminent thinkers. Over the years, a large number of false proofs were produced, purporting to deduce the fifth postulate from the first four. Some were even accepted as definitive, until an error was eventually found.

How to prove a negative? Build a new world

While some mathematicians believed there should be a proof of the parallel postulate from the first four, others disagreed. They believed that Euclid had not missed a trick, and that the fifth postulate really is *logically independent* of the others. The difficulty for these theorists was: how could they ever hope to demonstrate the truth of their belief? If someone was to provide a proof of the parallel postulate, that would settle the matter in one direction. But how could anyone try to show that no such proof can ever exist?

The answer arrived in the 19th century, when several mathematicians separately found the solution. If they could build a new system of geometry, which satisfied the first four of Euclid's postulates, but failed to satisfy the fifth, then it could not be an automatic consequence of the others, but must indeed be independent. Such a system would also be of great interest in its own right. What could it look like? Certainly the ordinary system of points and straight lines on a page satisfies the parallel postulate, so mathematicians needed to broaden their horizons.

In search of new worlds

Words can be misleading. For centuries people were trapped by the language of 'straight lines'. The conceptual leap that was needed was to stop thinking of 'straight lines' as we are familiar with them, and instead look for any system of lines that satisfies the first four postulates. These will then be the 'straight lines' of the new geometry, whatever they may look like to the human eye. When this breakthrough finally came, it was made not just once, but simultaneously by the mathematicians Carl Friedrich Gauss, Nikolai Lobachevsky and János Bolyai.

The system they discovered is today known as *hyperbolic geometry*. There are several different ways to visualize this new space, the most popular of which was later described by Eugenio Beltrami in 1868. It begins with an ordinary circular disc: this will play the role of the whole plane. The 'straight lines' are now arcs of circles that cross the disc, but only those that hit the edge at right angles. Also included are diameters going straight across the disc.

The first of Euclid's postulates is satisfied by this arrangement, but what of the second? A stretch of 'straight line' cannot be extended beyond the edge of the disc, let alone indefinitely, so it seems to fail. The answer is that the entire notion of distance is changed in this situation. The distance that operates is not the ordinary Euclidean distance, but a new *hyperbolic distance*. According to this, the edge of the disc is infinitely far away from any point inside: however far you travel towards it, it always remains at the horizon. From the outside looking in, space seems to shrink as you get near the edge. Working with this notion of distance, postulates 2 and 3 are true.

Although the notion of distance is changed, that of angle remains the same as we are used to. So the fourth postulate is true. The fifth, however, now fails. If we have a 'straight line' and a point inside the disc not on the line, there are in fact infinitely many 'straight lines' that pass through the point, but never cross the original line, and therefore are parallel to it. Bolyai's strange new universe finally resolved the status of the parallel postulate: it is logically independent of the others.

● The shape of the Earth and the universe

If all this fretting about axioms and strange new geometries seems of arcane interest at best, think again. Since its discovery, hyperbolic geometry has assumed a pivotal role in modern science, equal to that of its Euclidean cousin. After all, what reason is there to assume that the universe we inhabit is Euclidean? Gauss suspected that it may not be, and thanks to the later relativity theory of Albert Einstein and Hermann Minkowski, we know that the geometry of motion is actually hyperbolic rather than Euclidean (see *How to slow time*).

'Out of
nothing I
have created
a strange
new
universe.'

JÁNOS BOLYAI

The discovery of hyperbolic geometry established that non-Euclidean forms of geometry could exist. So, what other possibilities might there be? The answer is that we live on a third form of geometry. A 'straight line' can be defined as the shortest path between two points. In ordinary Euclidean space, this matches up with the ordinary notion of a straight line. In hyperbolic space it produces the curved arcs that cross the disc.

But what is it on planet Earth? If we take two points on a sphere, the shortest path between them is a straight tunnel between them. But if we want to remain on the surface, the answer is given by a stretch of *great circle*. These are the largest circles the sphere can contain, and divide it into two equal hemispheres.

Where does this new notion of a straight line leave the parallel postulate? In Euclidean geometry, if we start with a straight line and a point off it, there is exactly one line parallel to the starting line that passes through the point. (That is exactly the parallel postulate.) On the hyperbolic disc, there are infinitely many lines passing through the point, all parallel to the starting line. But in *spherical geometry*, there are none at all. Any two 'straight lines' (that is great circles) will eventually meet.

The disc model of hyperbolic geometry

A trinity of geometries

In all the years that people were investigating the parallel postulate, why did no-one notice that our very planet is non-Euclidean? The answer is that Euclid's third postulate does not hold on a sphere in its original form, since the great circles represent the maximum size of circle we can draw. So, in the 19th century it finally became necessary to update Euclid's postulates, to act as a modern foundation for the subject. With this done, there are now known to be three essentially different forms of geometry: hyperbolic, Euclidean and spherical.

Nowadays mathematicians no longer find these as strange as they first seemed. In each setting, our concerns are still the traditional questions of geometry, such as how triangles behave. In Euclidean geometry, every schoolchild knows that the angles in any triangle add up to 180°. In hyperbolic geometry, they add up to less, giving the triangles a 'thin' appearance. In spherical geometry however, the angles add up to more than 180°, and look 'fat'. Indeed, here we can draw triangles that contain three right angles. On the planet Earth, one such has its corners in Gabon in West Africa and Sumatra in Indonesia (both on the equator at 13° and 103° east respectively), and at the north pole.

19 How to unknot your DNA

- Knot theory
- Knot tables and classifications
- Left and right-handed knots
- Polynomials and invariants

In the late 19th century, a fundamental question of physics was coming to a head. It was the atom: what did this tiny speck of matter look like? At the time, there was a popular theory of aether, a universal substance that is everywhere, passing through everything. The physicist and mathematician Lord Kelvin had a radical idea about how matter emerges from aether. Atoms, said Kelvin, are little whirlpools in the aether, knotted in loops.

Lord Kelvin made several major contributions to mathematics and physics (see *How to build the perfect beehive*, for example), but it is fair to say that the theory of vortex atoms was not his greatest achievement. The theory of aether was knocked down shortly afterwards, and the atom would come to be more accurately described by the scientists Ernest Rutherford and Niels Bohr in the coming century.

Although Kelvin's theory of vortex atoms was wrong, it nevertheless left a scientific legacy that survives to this day. In his model, to tell two different elements apart, it was necessary to tell apart their corresponding knots. Over the 20th century, knots would again come to occupy an important position in science. The question of telling them apart would prove astoundingly difficult to solve. In a surprise twist to the tale, partly vindicating Kelvin, it would turn out to be a critical question in quantum physics.

No loose ends

A mathematical *knot* is the same as an ordinary piece of knotted string, with one slight change: after the string has been knotted, the ends are fused together, resulting in a loosely knotted loop. (A good way to model knots at home is to use an electric extension cable, plugging the two ends together after you've knotted it.)

Knot theory forms part of the mathematical subject of topology (see *How to find all the holes in the Universe*). Here it is not the precise geometric details of the shape that are important. If you pull or stretch the loop, to a topologist it is still essentially the same knot, as long as you do not cut or glue it. This relaxed approach causes a difficulty: by pulling the knot around, it is possible to make it *look* utterly different. So, when presented with two knots, it is by no means obvious whether or not they are actually the same in disguise. The ultimate aim of knot theory is to find a technique that will answer this question with complete accuracy every time. This turns out to be a question of exceptional difficulty, cutting right to the heart of the 3-dimensional space we live in.

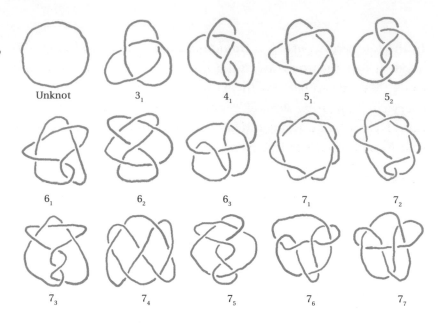

A knot table of up to seven crossings

Unknot 3_1 4_1 5_1 5_2

6_1 6_2 6_3 7_1 7_2

7_3 7_4 7_5 7_6 7_7

The simplest of all knots is known as the 'unknot'. It is just an unknotted loop of string. At least, that is what it usually looks like, but even the unknot can adopt clever disguises and *appear* thoroughly knotted, although of course this is an illusion.

One problem we might like to solve then, when presented with a tangled mess, is to determine whether it is actually knotted at all, or is really the unknot in disguise. This is the so-called *unknotting problem*, and is the simplest case of the more general question: how can we tell whether or not two given knots are really the same?

The unknot in disguise

A periodic table of knot theory

In the mid 19th century, Lord Kelvin's work sparked the first wave of interest in knot theory. Teaming up with Peter Tait, he started to list knots, based on the number of crossings that the picture of the knot contained. The unknot is the only knot with zero crossings. Knots with one or two crossings can all be untwisted to reveal the unknot. The first non-trivial knot is therefore the 3-crossing *trefoil* knot. Then the *figure-of-eight knot* is the only knot with four crossings, but there are two with five crossings, three with six, seven with seven and 21 with eight. As the number of crossings increases, the number of knots grows very fast. In the final decade of the 19th century, Tait's small team of pioneering knot theorists managed to list all the knots up to 10 crossings, amounting to 250 different knots. In fact they had 251 knots in their tables, but had unwittingly included a duplicate. The two 10-crossing knots they had called 10_{161} and 10_{162} actually turned out to be the same, although this was not discovered until 1974, when the amateur knot theorist Kenneth Perko spotted the mistake. The fact that the experts had overlooked this error for the best part of a century illustrates the high level of difficulty of telling knots apart.

These early knot tables formed the backbone of the subject, and were only substantially improved once the computing power of the late 20th century could be exploited. In 1998, Hoste, Thistlethwaite and Weeks published a paper 'The first 1,701,936 knots', which completely classified all knots up to 16 crossings.

The Jones polynomial

In 1984, Vaughan Jones was a researcher in a completely different field of mathematics, when he made an unexpected breakthrough in the analysis of mathematical knots. Hidden in his work was a new way to tell knots apart.

A trefoil knot with writhe +3

The starting point was to calculate a number associated to the knot, called the *writhe*, as follows. There are two directions we could travel around the knot. First we fix one, by placing a consistent set of arrows going around the string. The crucial points on the diagram are now the crossings. We assign each crossing either a number +1 or −1, according to the pictured rule. Then we add up all these numbers to get the *writhe* of the knot. It is *almost* true to say that the writhe is uniquely defined for the knot, but of course we can always add an extra twist, thereby changing the writhe. However, this is the only way to change the writhe.

The Jones polynomial is essentially a way to correct this by introducing a little algebra. Starting with a knot such as the trefoil knot (pictured), Jones found a method to derive an algebraic expression from it, in this case $x + x^3 − x^4$. The result has a very useful property indeed, namely that, however much you twist and manipulate the knot before calculating, you always finally arrive at the same polynomial at the end. This is exactly what knot theorists want: when you start with two knots and get two different polynomials at the end (say one produces $x + x^3 − x^4$ and the other $x^2 − x + 1 − x^{-1} + x^{-2}$) then you know for sure that the two knots really are fundamentally different. No amount of pulling around would ever pull one into the other.

The idea was not new. A similar technique had been developed by James Alexander in 1923. However, there were several ways in which Jones' polynomial was a major advance on Alexander's; indeed it was the first major theoretical breakthrough in 60 years.

Lefties and righties

*Left-handed
trefoil knot*

*Right-handed
trefoil knot*

Start with a knot, such as the trefoil knot. Suppose we now take the mirror image of the knot. Reflecting something in a mirror does not qualify as pulling or stretching it, so the question is, is this reflection a new knot, or is it really the same as the original? Certainly it looks different, but might it be possible to pull one in to the shape of the other? In this case the answer is no, as a little experimentation will show. However, if we start with the figure-of-eight knot (pictured below), this time the mirror image is equivalent to the original, although this is not immediately obvious. Knots like this are called *achiral*. For the first time, the Jones polynomial provided a method for deciding whether a knot is *chiral* like the trefoil (meaning that it comes in two versions, a left-handed and right-handed version), or achiral such as the figure-of-eight knot, where the two versions are actually the same. Chirality is an important property in chemistry. Some molecules come in both left-handed and right-handed versions, each with subtly different chemical properties.

The knots of life and death

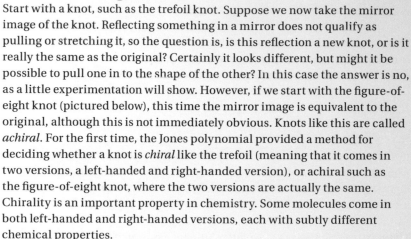

*Left-handed
figure of eight*

*Right-handed
figure of eight*

Knots appear in many places in the natural world. One such place is inside you. In each cell of the human body lives a large amount of DNA, the molecules that store the information for programming a human body. As James Watson and Francis Crick discovered in 1953, a DNA molecule has the form of a double helix (corkscrew). Each cell nucleus contains around three metres of DNA, which drifts around in long strands. As anyone who has untangled a box of wires will attest, the tendency in such a situation is for the strands to get horribly knotted with each other. This is a mixed blessing: as was discovered in 2007 by Richard Deibler and colleagues. Knots are toxic: knotted DNA will usually lead to the death of the whole cell. At the same time, knots in DNA are also a driver of genetic mutation, the engine of evolution.

To combat the tendency for DNA to get knotted, each cell contains an army of enzymes that work on the DNA. As well as carrying out essential repairs and copying and pasting the genetic code during cell reproduction, these enzymes fight against knots, carefully untangling any sections that have become knotted.

A computer program is useless without a computer to run it on. The same goes for a genetic code. The cellular enzymes form the first level of the apparatus that converts genetic information into action. Therefore understanding how they work is a major topic in biochemistry. In recent years, mathematics has contributed a great deal to this work. To determine exactly what an enzyme does, scientists had the idea of letting the enzyme work on a closed loop of DNA, instead of a strand with free ends. After the enzyme had chopped and reassembled the DNA, the result was a knotted loop. By studying this knot using the Jones polynomial, scientists were able to deduce what action the enzyme must have performed.

Kelvin's vortices: right after all?

The Jones polynomial was a breakthrough: it could immediately distinguish between knots that were previously almost impossible to tell apart. In particular, it was excellent at detecting chirality. However, it was still not perfect. There are still some knots that are genuinely different, but which have the same Jones polynomial. Jones' work provoked the second wave of interest in knot theory, and several ways to improve upon his polynomial were found. These new *knot invariants* as they are called were able to distinguish between more knots than even the Jones polynomial, but the search for a perfect invariant that can distinguish between any two knots is a major ongoing project in mathematics today.

This search for more powerful knot invariants has converged in a highly unexpected way with fundamental questions of quantum physics. Jones' invariant was a fairly simple mathematical object called a polynomial, just some different powers of x added together. New invariants, such as the *Khovanov homology* discovered in 2000, are of considerably more abstract and technical types. They are formed by viewing the Jones polynomial as a shadow of a larger, more complex object capturing a huge amount of information about a knot.

The idea is to rebuild the larger object from its shadow, by a process of *categorification*. Physicists had been working along very similar lines. One of the major goals of modern physics is to reunite Einstein's theory of relativity (see *How to slow time*), which describes gravity across the universe, with quantum mechanics, which describes the behaviour of individual particles. Perhaps these two are separate shadows of the same ultimate mathematical model of the universe? In the attempt to build such a model, the mathematics of knots has returned to prominence, over a century after Kelvin's knotted atoms fell out of fashion.

'Oh! What a tangled web we weave, When first we practice to deceive!'

Walter Scott, *Marmion*

20 How to find all the holes in the universe

- Shapes without holes
- 2-dimensional surfaces and 3-dimensional manifolds
- Henri Poincaré's bold conjecture
- Grigori Perelman's fluid proof

We might expect that counting holes is a job reserved for local road-inspectors, but it is also the basis of an entire mathematical discipline, called topology. A particularly important question is: which shapes have no holes at all? This was answered in one of the triumphs of modern mathematics: the Poincaré conjecture.

Why a square is really a circle

The usual approach to geometry is to study the fine details of shapes such as circles and triangles. A geometer might analyze the precise amount by which a circle curves, or make calculations about the angles inside a triangle. But, to a topologist, triangles, squares and circles are all the same thing.

Topology, or rubber-sheet geometry as it sometimes known, is concerned with the properties of shapes that are unaffected by any amount of pulling or twisting. Since a square can be pulled into the shape of a circle, in topological terms it *is* a circle. (This is nothing to do with the ancient problem of *squaring the circle*.)

At this stage, you might object that anything can be pulled into the shape of anything else, if you set about it with enough violence. However, topology has two inviolable rules. While you can stretch or bend a shape, you may never cut or glue it. So the number '8' cannot be pulled into the shape of the number '0', because cutting is forbidden. Neither can the letter 'j' be pulled into the shape of an 's' since its two parts may not be glued together.

Is the Earth flat, round or donut-shaped?

One of the greatest scientific discoveries of all time was surely the revelation that the Earth is round, rather than flat. This was first hypothesized by early Greek thinkers such as Pythagoras, and ultimately confirmed with the dawn of the age of nautical exploration, notably Christopher Columbus' attempt to journey westward to the East Indies.

It seems surprising that this was ever at issue, given that a sphere and a flat plane are so utterly different. How could anyone have ever confused the two? The answer is that they are only different when viewed from afar. From the perspective of a creature on the surface, every patch of sphere looks very much like a patch of plane.

This is the definition of what mathematicians call a *surface*: a 2-dimensional object where every region looks like a plane. The sphere is one example, but

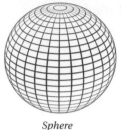
Sphere

there are others too: the donut-shaped torus, or the double torus with its two holes. It was only once the Earth had been comprehensively mapped that we could be sure that we did live on such a surface.

There are wilder possibilities too, such as the *Klein bottle*. Here, regions of plane are patched together in a coherent way, but the result cannot be represented in 3-dimensional space without colliding through itself.

Torus

The sphere, torus, and Klein bottle are all *closed surfaces*, meaning that they come in one piece, have no edges, but nevertheless have a finite area (unlike an infinite plane that continues for ever in every direction). Since the mid 19th century, mathematicians have had a good grasp of all possible closed surfaces.

Double torus

Topological fishing: how to catch a hole

How can you tell whether a shape contains a hole? This question is one to which mathematicians of the 20th and 21st centuries have devoted an enormous amount of thought. The starting point is to take a sphere and draw a loop on it. Then, by slowly tightening the loop, you can gradually shrink it away to nothing. When you try the same trick on a torus, if the loop encircles the hole, then it cannot be shrunk away: it is stuck. Contractible loops, then, are excellent at detecting holes (see diagrams on pages 126–127).

Klein bottle

Henri Poincaré saw that the sphere is the only surface where every loop can be shrunk away to nothing. All others have holes of some type. (Of course, a cube has no holes either. But remember that, to a topologist, a cube is a sphere.) At the dawn of the 20th century, Poincaré revisited this question, stepping up a dimension. Instead of working with 2-dimensional surfaces, he considered 3-dimensional objects called *manifolds*.

The view from the fourth dimension

1-dimensional space is a straight line extending infinitely in both directions, 2-dimensional space is an infinite plane; 3-dimensional space is where we live (or seem to), with freedom to move left, right, up, down, forwards and backwards. From these spaces more exciting shapes can be built. Just as a circle looks like a line bent round, and a surface is patched together from pieces of curved plane, so a *3D manifold* is a shape patched together from pieces of 3-dimensional space.

The space we inhabit is a 3D manifold. We may imagine that it is flat, extending forever in every possible direction. But we should beware the example of the ancient flat-earthers, who made exactly the same assumption. Mathematically, there are plenty of 3D manifolds our universe could be, which curve round on themselves just as the sphere or torus do. Admittedly, the limitations of the human mind make these shapes difficult to imagine. The problem is similar to that of the curvature of the Earth, namely that it is only visible when you view it from further away. Similarly, a 3D manifold looks like flat 3-dimensional space from the inside. To see its overall shape, you need to view it from the outside, from a new perspective along the *fourth dimension*.

The fourth dimension is something that our brains are not wired to visualize. Happily, however, in mathematics we have powerful techniques for analyzing higher-dimensional shapes, without having to tie our minds in knots trying to picture them.

Hyperspheres and hyperdonuts

Many 3D manifolds are weird places, higher dimensional analogues of the torus and Klein bottle, and some even more alien than these. However, there is one that is comparatively straightforward. The *3D sphere* is the 3-dimensional equivalent of the usual 2-dimensional sphere.

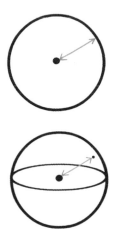

Although we can never hope to visualize this shape (it curves around in the fourth dimension) we can at least understand its structure. Put a dot on a piece of paper. Now mark all the places on the page that are 1 cm away from that spot. The shape that emerges is a circle. The same strategy in 3-dimensional space produces an ordinary sphere, again defined as the set of all points a fixed distance away from the centre. This line of thought works in higher dimensions too. A *3D sphere* is defined as the set of points a fixed distance away from the centre, in 4-dimensional space. In each case, topologists then allow the result to be twisted into more exciting configurations.

Spheres in two and three dimensions

Mathematics' hole in the heart

How can we hope to understand 3D manifolds, which we can never hope to picture? A valuable technique is to generalize and extrapolate from more familiar scenarios, in this case 2-dimensional surfaces.

The question Henri Poincaré addressed in the early 20th century was: which are the 3D manifolds with no holes? Poincaré believed that the answer should be exactly the same as for surfaces: only the sphere should qualify. This time, however, it is not the familiar 2-dimensional sphere, but its big brother, the 3D sphere.

Poincaré's idea carried the comforting message that these higher dimensional manifolds are not completely alien. Poincaré was a truly great mathematician, often cited as the last with mastery of the whole subject. Unfortunately, though, neither he nor any of his contemporaries were able to prove his hunch correct. It remained possible that some mysterious new shape could be found, not a sphere, yet containing no holes. The *Poincaré conjecture*, as it became known, came to occupy a central position in mathematics. Its refusal to surrender to generations of mathematicians was a reminder of our continued inability to tame the fourth dimension.

Hamilton's shape-shifters

Despite the attentions of the greatest minds of the 20th century, the Poincaré conjecture remained unresolved at the dawn of the 21st. Its lure increased even further when it was selected as one of the Clay Mathematics Institute's Millennium problems. This meant that a $1,000,000 prize now awaited anyone able to prove it. As it happened, the solution was nearer than anyone expected.

Many things in the world *flow*: water, heat and money are three examples. Such phenomena present a challenge for mathematicians, as flow is difficult to understand. Although many sophisticated techniques have been developed, fundamental questions remain unanswered today. Fascinating as it is, no-one expected that this topic should have any bearing on such a classical question as the Poincaré conjecture. That is, until Richard Hamilton came up with the idea of *Ricci flow* in 1981. The idea was to treat curvature not as something fixed, but as something able to evolve over time, just as heat flows from a warmer patch to a cooler.

A contractible loop on a sphere and on a torus

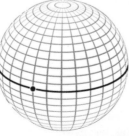

How to find all the holes in the Universe

Hamilton had an idea about how to prove the Poincaré conjecture. Starting with a 3D manifold, he could set the curvature flowing. Then, if the original manifold had no holes, he believed that it should eventually settle into the shape of a sphere. It was a beautiful idea, but the obstacles were formidable. The major difficulty was that the flow sometimes led to *singularities*, such as a piece of the shape getting pinched off.

Perelman wields his scalpel

In 2002, the Russian mathematician Gregori Perelman announced an astonishing discovery. He argued that singularities were not obstacles to a proof, after all. Rather, they were the keys to it. Perelman's idea was to let the flow run until a singularity appeared. Then he paused the flow and, using a new technique called *surgery theory*, he cut out an area around the singularity. Once removed, the excised part had a smooth shape, easy to understand. Then he restarted the flow on what remained. After a number of surgeries had been performed, the manifold had been chopped up into pieces. Each piece, Perelman showed, must be one of eight fixed types.

Mathematicians around the world set about scrutinizing Perelman's work. By 2006, it was accepted that Perelman had not only proved the Poincaré conjecture, he had delivered even more. He had given a complete account of all the possibilities for any 3D manifold. Such a list had been predicted, in detail, by William Thurston in 1982. His *geometrization conjecture* went so far beyond the Poincaré conjecture that many mathematicians despaired of ever seeing it proved.

The reclusive Perelman declined the prize money, and even turned down a Fields medal, mathematics' equivalent of a Nobel Prize. But his work, building on that of Richard Hamilton, and following the visions of Henri Poincaré and William Thurston, means that humans can now legitimately claim to have conquered the fourth dimension.

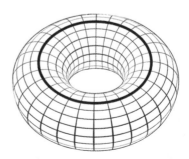

21 How to feel at home in five dimensions

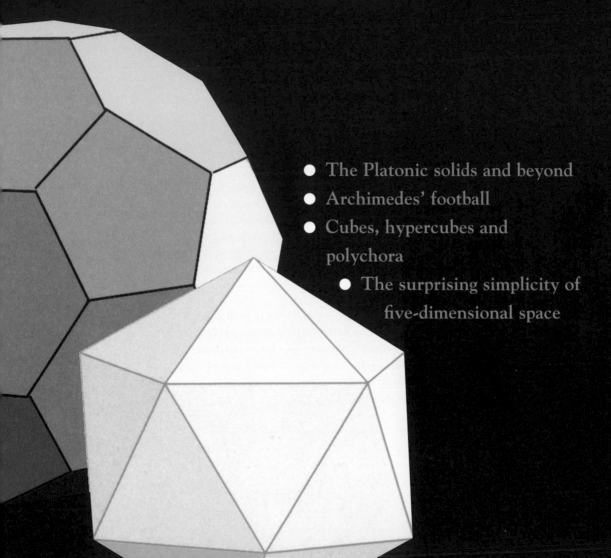

- The Platonic solids and beyond
- Archimedes' football
- Cubes, hypercubes and polychora
- The surprising simplicity of five-dimensional space

The Great Pyramid in Giza has a square base and four triangular sides. A paperback book has six sides, each rectangular. A soccer ball has 16 pentagonal faces and 20 hexagonal ones. These are all what mathematicians call polyhedra: shapes built from straight edges and flat faces, meeting at corners. The study of polyhedra is one of the longest-running themes in mathematics, and the source of some of the most elegant of all theorems. When lifted into higher dimensions, the exotic cousins of the polyhedra form the key to understanding the structure of dimensions far beyond our own.

Polyhedra are the 3-dimensional counterparts of *polygons*, 2-dimensional shapes such as triangles, rectangles and pentagons. The most symmetrical of these shapes are those whose angles are all equal and whose sides are all the same length. These are the *regular polygons*. The first is the equilateral triangle with its three equal sides. Then come the square, the regular pentagon, the regular hexagon, and so on. In fact, a regular polygon can be built with any number of sides starting with three. When the geometers of the ancient world tried to lift this reasoning into three dimensions, they found the situation more complicated. By considering rather subtle notions of symmetry and regularity, several wonderful families of beautiful shapes blossomed forth, beginning with the most famous of all: the *Platonic solids*.

The shapes of perfection

One of the most symmetrical of all polyhedra is the cube. What is it that makes this shape so special? Firstly, each of its six faces is identical to every other. What is more, each face is a square, itself a highly symmetrical shape: a regular polygon where all the edges are the same length, and all angles identical. The Greek philosopher Plato took these two conditions as the starting point for a complete account of all such polyhedra. He proved what is known as a *classification*, the very highest rank of mathematical theorem. Plato's work amounts to a complete list of all possible shapes that satisfy his two criteria: that every face is a regular polygon, and that all are identical. He proved that there are only five such shapes (see page 130). They are:

1. The tetrahedron, which has four faces, each an equilateral triangle. The tetrahedron is a triangular based pyramid.
2. The cube, with its six square faces meeting together in threes.
3. The octahedron, which consists of eight equilateral triangular faces. (It looks like two square-based pyramids glued together at their bases.)
4. The dodecahedron, with 12 pentagonal faces.
5. The icosahedron, in which 20 equilateral triangles meet together in fives.

Tetrahedron

Octahedron

Cube

Dodecahedron

Icosahedron

> *'The knowledge at which geometry aims is the knowledge of the eternal.'*
>
> PLATO, *The Republic*

These five Platonic solids were a source of awe in the ancient world. They took on a divine significance, coming to represent the perfection and purity of the geometrical realm. The greatest geometers of the classical era, Pythagoras, Euclid, Archimedes and others, devoted a great deal of time to their study. Plato himself believed that these shapes represented the very foundations of reality, with the classical elements of earth, air, fire and water taking the forms of the cube, octahedron, tetrahedron and icosahedron, respectively. The cosmos itself was held to be a dodecahedron. This idea of the cosmic significance of the Platonic solids was long lived, being revived by Johannes Kepler. It was a grand thought, but there is no evidence that these shapes are writ large across our universe.

● Archimedes' football

The Platonic solids are built from regular polygons: shapes such as equilateral triangles, squares and regular pentagons. What is more, in a Platonic solid each such face is identical to every other.

To extend the list of polyhedra, we could drop this second requirement, but still maintain a high level of symmetry. This was the approach adopted by one of the greatest of all scientists, Archimedes of Syracuse.

The definition of an Archimedean solid begins as the definition of a Platonic solid does, namely that the faces must all be regular polygons. This time, however, they need not all be the same. Some faces may be square, others triangular, pentagonal, and so on. However, all the edges on the shape must be the same length. To preserve symmetry, Archimedes made one final requirement: the arrangement of faces that meet at each vertex must be the same as at every other.

From this rather abstract definition, Archimedes conjured up 13 beautiful new shapes: the Archimedean solids. As with the Platonic solids, this is a classification: a complete list of all the shapes satisfying the criteria, except this time the 13 shapes alone are not quite enough. There are also two infinite families that satisfy the conditions. Start with two identical regular polygons, say two hexagons. If we connect these with a ring of squares, we get a hexagonal *prism*. Instead, if we twist the two hexagons out of sync, and

How to feel at home in five dimensions

connect them with a ring of equilateral triangles, we get a hexagonal *antiprism*. Prisms and antiprisms can exist for any polygons: heptagons, octagons, and so on, and so there are infinitely many of them.

The first among the Archimedean solids is the *cuboctahedron*, built from six squares and eight triangles. Five of Archimedes' shapes are formed by truncating a Platonic solid, that is to say cutting its corners off. When you truncate a cube, for example, the six squares are trimmed into six octagons, and a triangle appears at each of the eight corners.

Soccer ball or truncated icosahedron

One of the most familiar of the Archimedean solids is the truncated icosahedron, better known under its other name of a football (or soccer ball). Perhaps the most complex Archimedean solid also comes with the most formidable name: the *great rhombicosidodecahedron*, which has 30 square faces, 20 hexagonal faces and 12 decagonal (that is, ten-sided) faces.

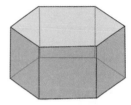

Hexagonal prism

What Plato missed

One of the critical criteria for geometry is the notion of convexity. It is a deceptively simple idea. The general idea is that a convex shape should not have any parts that stick out too far from the body of the shape.

Suppose you pick two points inside a circle, and join them with a straight line. Whichever points you pick, the resulting line will lie completely inside the circle (see *How to arrange the perfect dinner party*). The same goes for a triangle, or a square. However, if you work with a star shape (or pentagram) and pick points near the tips of adjacent arms, then the resulting line will pass outside the shape.

Hexagonal antiprism

This means that the circle, triangle and square are *convex* shapes, whereas the pentagram is *non-convex*. Exactly the same definition applies to 3-dimensional polyhedra. All the Platonic and Archimedean solids are convex. In the 17th century, Johannes Kepler recognized the possibility of non-convex shapes that might also satisfy the definition for a Platonic solid. He found two new shapes, and in 1809 Louis Poinsot found another two. These four intricate new Kepler–Poinsot polyhedra are the non-convex equivalents of Plato's famous solids. There is another cost to these, however. As well as being non-convex, the edges of the pentagram pass through each other, resulting in *false vertices*.

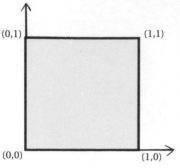

(0,1)   (1,1)

(0,0)   (1,0)

(0,0,1)   (1,0,1)
    (1,1,1)
(0,1,1)
        (0,0,0)
                (1,0,0)
(0,1,0)
        (1,1,0)

Coordinates of a square and a cube

● Squares, cubes and hypercubes

The most familiar of the Platonic solids is the cube, itself a higher dimensional equivalent of the 2-dimensional square. The system of coordinates makes this easy to understand. A square is completely captured by defining the coordinates of its corners: $(0, 0)$, $(0, 1)$, $(1, 0)$, $(1, 1)$.

Similarly, a cube can be defined by the coordinates in three dimensions of its eight corners: $(0, 0, 0)$, $(0, 0, 1)$, $(0, 1, 0)$, $(0, 1, 1)$, $(1, 0, 0)$, $(1, 0, 1)$, $(1, 1, 0)$, $(1, 1, 1)$. Even with no understanding of the geometry here, it is clear what the pattern is: the corners of the square are represented by every possible pairing of 0s and 1s. Similarly, the corners of the cube are given by every possible triplet of 0s and 1s. This tells us how to define a *hypercube*: the 4-dimensional analogue of the cube. It is the shape that has corners at: $(0, 0, 0, 0)$, $(0, 0, 0, 1)$, $(0, 0, 1, 0)$, $(0, 0, 1, 1)$, $(0, 1, 0, 0)$, $(0, 1, 0, 1)$, $(0, 1, 1, 0)$, $(0, 1, 1, 1)$, $(1, 0, 0, 0)$, $(1, 0, 0, 1)$, $(1, 0, 1, 0)$, $(1, 0, 1, 1)$, $(1, 1, 0, 0)$, $(1, 1, 0, 1)$, $(1, 1, 1, 0)$, $(1, 1, 1, 1)$.

Of course we do not need to stop here: there are hypercubes in five, six, seven, eight and as many dimensions as we like. Visualizing these shapes is not directly possible, but mathematics nevertheless gives us a way to start analyzing them. What is more it is clear that the hypercube is built from cubes, just as the cube is built from squares. This prompts a fascinating question: what are the 4-dimensional equivalents of the Platonic solids? These are the so-called *Platonic polychora*.

● Seeing in four dimensions

In 1852, Ludwig Schläfli accomplished in four dimensions what Plato had done in three. He provided a complete classification of the regular convex polychora. He found that there are six:
- the *pentachoron* (or *simplex*), built from five tetrahedra
- the *hypercube*, built from eight cubes
- the *orthoplex*, built from 16 tetrahedra
- the *octaplex*, built from 24 octahedra
- the *hecatonicosachoron*, built from 120 dodecahedra
- the *hexacosichoron*, built from 600 tetrahedra.

The pentachoron, hypercube, orthoplex, hecatonicosachoron, and hexacosichoron are analogues of the ordinary Platonic solids. The octaplex, however, was a brand new shape, with no 3-dimensional equivalent.

Exploring one million-dimensional space

What happens if we lift Plato's idea into higher dimensions still? With five Platonic solids in three dimensions, and six in four dimensions, we might expect to see ever more complicated shapes in higher dimensions. How could we ever hope to understand million-dimensional space, for example?

Some trends are identifiable. We can certainly build a hypercube in every dimension, directly from the patterns of numbers within the coordinates. Exactly the same thing is true for the octahedron, which has its corners at $(0, 0, 1)$, $(0, 0, -1)$, $(0, 1, 0)$, $(0, -1, 0)$, $(1, 0, 0)$ and $(-1, 0, 0)$. Its 4-dimensional equivalent is the *orthoplex*, which has its corners at $(0, 0, 0, 1)$, $(0, 0, 0, -1)$, $(0, 0, 1, 0)$, $(0, 0, -1, 0)$, $(0, 1, 0, 0)$, $(0, -1, 0, 0)$, $(1, 0, 0, 0)$ and $(-1, 0, 0, 0)$.

A hypercube and orthoplex will certainly exist in every dimension, and the same is true for the tetrahedron or *simplex*. An equilateral triangle (2-simplex) is formed by three points that are equal distances apart. Similarly, a tetrahedron (3-simplex) is formed by four points that are equal distances apart. Meanwhile, the pentachoron (or 4-simplex) is defined by five points that are all the same distance apart; it lives in four dimensions. This process extends to produce simplices in all higher dimensions.

So there are three types of shape that will feature in every dimension: the simplex, hypercube and orthoplex. The trickier ones to understand are those with no such easily discernible pattern, the likes of the dodecahedron and octaplex. When we look at five dimensions and higher, Schläfli realized that something truly extraordinary occurs: all of these complicated shapes disappear. The only Platonic shapes that are left are the three we can easily understand.

This illustrates a fact that very few people appreciate. It is common to see higher dimensions as weird and mysterious places compared with our comfortable home in 3-dimensional space (or 4-dimensional space when you include time). Actually, from a mathematical perspective, this has it entirely backwards. It is 3- and 4-dimensional space that are cramped and difficult to understand. With the extra room for manoeuvre that higher dimensions provide, these difficulties evaporate, and geometry works much more cleanly. For example, in three dimensions one of the most difficult questions in geometry is to analyze the different ways that a loop of string can be knotted (see *How to unknot your DNA*). However, if we pose the same question in higher dimensions, it is trivial: any string can automatically be unknotted.

'Dimensions are limitless; time is endless. Conditions are not invariable; terms are not final. Thus the wise man looks into space and does not regard the small as too little, nor the great as too much; for he knows there is no limit to dimension.'

CHUANG TZU, *Neither Great nor Small*

22 How to design the perfect pattern

- The 17 wallpaper patterns
- Symmetries – translations, rotations, reflections and glides
- Crystals and crystallography
- The past masters of symmetry

According to Georges Polti, there are only 36 possible plotlines that any story can follow. How can this possibly be true? It is easy to imagine endless new tales about bereaved bee-keepers stuck in traffic, or homesick time-travellers taking solace in virtual reality; the number of possible scenarios, characters and storylines is self-evidently infinite.

What Polti means is that, when you strip away the superficial details of the characters, the time and place of the action, and focus on the basic constituents of the underlying dramatic situation, only 36 possibilities remain. One of them involves someone set on revenge for an earlier crime (as in *James Bond, Licence to Kill*); another is the tale of a person who falls in love with an enemy (such as in *Romeo and Juliet*).

If Polti is right, his analysis must occur at a suitable level of abstraction, according to which certain stories are deemed to be 'essentially the same'. Of course, people will argue forever about whether or not his thesis is correct, because literary theorists do not have access to the benefits of mathematical proof. If it is right, Polti's work amounts to what mathematicians would call a *classification* of storylines.

In 1891, Evgraf Fedorov proved something very similar for patterns, such as those found on wallpaper (either of the domestic or computerized variety), carpets, clothing fabric, the mosaics of ancient Rome and the Islamic world, and indeed innumerable other situations. Of course, in any of these we can have pictures of hummingbirds, or battleships, or any number of abstract designs, and so on. It is not sensible to attempt a complete list. But when one strips away the superficial details, Fedorov showed that there are exactly 17 possible underlying structures that any pattern can have. Today, these are known as the 17 *wallpaper groups*. Unlike Polti's storylines, there is no continuing debate here, as Fedorov was able to provide a watertight mathematical proof of his classification.

'*Nature uses only the longest threads to weave her patterns, so that each small piece of her fabric reveals the organization of the entire tapestry.*'

Richard Feynman

Repeating itself, and repeating itself, and repeating itself, and . . .

What is the difference between a *pattern* and a *picture*? The answer, for present purposes at least, is that a pattern repeats itself, whereas a picture need not. To be more precise, if you slide the pattern rightwards you will eventually come to a point where it looks exactly the same as when you started. The patterns Fedorov considered repeat themselves in two different ways: first if you slide it right (or left), and secondly if you slide it up (or down). One consequence is that the pattern is essentially infinite. Of course, on any particular wall or carpet, only a finite area will be covered, but the rules of the pattern make it easy to extend as far as you like in any direction.

This property of looking the same when you slide it along is what mathematicians call *translational symmetry*. This is the beginning of the classification, then: every pattern must, by definition, have two types of translational symmetry (up/down and left/right). In fact, the simplest patterns have only these symmetries and no others. These form the first of Fedorov's 17 types, denoted 'o'. If you take an ordinary, non-symmetric photograph and place copies of it on an infinite grid this will produce a pattern of type o. The other 16 types of pattern additionally have other symmetries, beginning with rotations.

Round and round and it always looks the same

Mathematicians use the word 'symmetry' in a slightly unusual way. Start with a single shape such as a square. A *symmetry* of this shape is an action you can perform that leaves it looking the same as before. For example, if we rotate our square by 90° (as if around a pin stuck in its centre), it looks the same as when we started. So this is one possible symmetry of the square. If you rotate it by 45° it looks different, so this is not a symmetry.

An important observation is that if we rotate the square by 90° four times, then we arrive back to the original position. So a square is said to have rotational symmetry *of order 4*. Similarly, a regular pentagon has rotational symmetry of order 5, and so on. The possible orders of rotation are a key ingredient to the wallpaper classification.

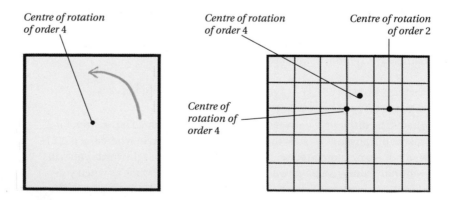

Centre of rotation of order 4

Centre of rotation of order 4

Centre of rotation of order 2

Centre of rotation of order 4

Just like individual shapes, infinite patterns can also have rotational symmetries. If we start with a simple pattern consisting of an infinite grid of squares and push a pin through the centre of one square, the whole pattern has rotational symmetry of order 4, exactly as the single square does.

However, something new happens with this pattern. If we push a pin through the crossroads where the sides of neighbouring squares meet, we can also rotate the pattern around there. This shows that while single shapes can only have one centre of rotation, patterns can have several. The crossroads point again forms a centre of rotation of order 4.

Actually the grid pattern has another type too: the middle of each edge is also a centre of rotation, but this time it has order 2. Only by rotating by 180° do we have a symmetry, not by rotating by 90°. This goes to show that, in a single pattern, not only are there different centres of rotation, but these can have different orders.

Diamonds and 229 other crystals

The observation that patterns can have several centres and that they can have different orders of symmetry might originally have been discouraging. It suggests that the number of possibilities may be complicated. However, there is a fact of immense importance that limits the possible rotations that a pattern can have. It goes by the name of the *crystallographic restriction theorem*.

This theorem says that patterns (which are assumed to have double translational symmetry) can only ever have rotational symmetry of order 2, 3, 4 or 6. This is an extremely useful result, as it dramatically reduces the possible symmetries that patterns can have. This fact allowed Fedorov to prove that a pattern that has only rotational symmetry must have one of four possible configurations:

632, 442, 333 and 2222

Each of these represents a type of pattern, and the individual digits list the orders of rotations. So 632 has three types of centre, with orders of 6, 3 and 2.

In fact, the crystallographic restriction theorem does not only apply to 2-dimensional patterns, but to 3-dimensional structures too. This makes it a fundamental tool for scientists who study the structures of physical materials such as diamond and graphite. This theorem dramatically limits the possible structures that any crystalline solid can have. Indeed, just as there are 17 possible wallpaper groups representing every possible 2-dimensional pattern, so the structures of symmetrical crystalline solids are classified by 230 *crystallographic groups*.

● Mirror, mirror on the wall

A single shape such as a square or a circle can have two types of symmetry. The first is rotational symmetry, where turning the shape leaves it looking the same. The second is *reflectional symmetry*, where reflecting it leaves it looking the same. If we place a mirror along a vertical line through the centre of a square, what we see in the mirror is exactly what lies behind it: the two sides are perfect reflected copies of each other. The same thing is true of the infinite square grid; this also has mirror lines through which it can be reflected, leaving it looking identical. In listing the 17 possible patterns, an asterisk is used to indicate that a pattern has reflections.

The grid has rotations of order 4, 4 and 2, but is not of the form 442 above, because it also has reflections. When we consider patterns with both reflections and rotations, something significant happens to the centres of rotation. As well as having different possible orders, centres of rotation now come in two essentially different forms: those that also lie on a mirror line, and those that do not. Those that do are called *kaleidoscopes*, while those that do not are called *gyrations*.

A pattern can have both gyrations and kaleidoscopes. For example, the illustrated pattern at top left has two types of rotation. Both are of order 3, but one is a gyration and the other is a kaleidoscope. We can illustrate this as 3*3, where the asterisk indicates that the pattern has reflections, the numbers before the asterisk list the gyrations, and the numbers after list the kaleidoscopes. So the infinite square grid has pattern *442, as all its rotations are kaleidoscopes. The

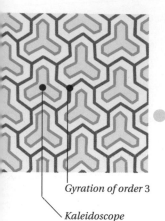

Gyration of order 3

Kaleidoscope of order 3

*Pattern 3*3*

*Pattern ***

Lines of reflection

possible patterns with both rotations and reflections are:

*632, *333, 3*3, *442, 4*2, 22*, *2222 and 2*22

There is also one possibility that has two parallel reflections but no rotations at all, which is denoted '**' (see page 138 foot).

The three ways to glide

As well as rotations and reflections (and of course translations), infinite patterns can have a fourth and final form of symmetry, which John Conway calls a *glide symmetry*. A glide is built by first reflecting the pattern across a line, and then sliding it parallel to that line. Although a glide manoeuvre is a combination of a reflection and translation, a pattern can have glide symmetries without having any reflectional symmetry, as the pattern below shows. We only count glides when they cannot be broken down into a reflectional and a translation symmetry. Glides are denoted by the letter x. There are three possibilities: patterns with two glides, patterns with one glide and one reflection perpendicular to it, and patterns with two gyrations of order 2 together with a glide. Respectively, these are denoted:

xx, *x and 22x

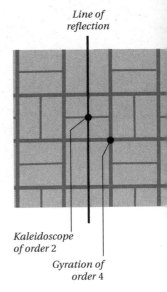

Line of reflection

Kaleidoscope of order 2

Gyration of order 4

*Pattern 4*2*

The past masters of symmetry

Altogether, the various possible combinations of symmetries produce 17 essentially different patterns one can draw (subject to the condition that they have double translational symmetry):

o, 632, 442, 333, 2222, *632, *333, 3*3, *442, 4*2,
22*, *2222, 2*22, **, xx, *x and 22x

These are the 17 wallpaper groups. Although the scientific analysis had to wait until Fedorov in the 19th century, all these patterns were investigated long before then, notably in the mosaics of Islamic temples. A game that mathematicians like to play is to find ancient examples of each of the patterns. It is even said that in the famous Alhambra palace in Spain all 17 patterns can be found.

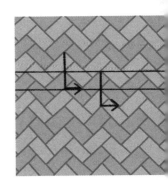

Pattern xx

We might ask, then, what the modern mathematics of symmetry has brought to the party. The answer is subtle but powerful. Thanks to Fedorov, we know that these are not merely 17 examples of symmetrical patterns, but are a complete catalogue, listing every pattern there can ever possibly be.

23 How to build the perfect beehive

- Dividing a space into cells
- Hexagons and honeycombs
- Kelvin's conjecture and Weaire–Phelan foam
- Packing circles and stacking spheres

Why is honeycomb hexagonal? One theory, popular in the ancient world, was that these six-sided shapes were somehow the easiest for the six-legged bees to navigate. This was never a convincing answer, and the real explanation requires more sophisticated mathematics than simply counting legs. Honeycomb is composed of tubes in which honey is stored. The tubes need to be identical and open at one end. The question is: what shape should the cross-section of the tubes have in order to use the least amount of wax?

The honeycomb question was the first in a line of geometrical questions that all involve maximizing the volumes of certain spaces, or fitting as many objects into a region as possible. These questions have two things in common. Firstly they are very natural and easy to ask. No deep geometrical concepts are needed; all the ideas can be understood by a child. The second thing they have in common is that they are exceptionally difficult to answer. The last of these geometrical questions to be solved, Kepler's sphere-packing conjecture, has proved so technically demanding that the proof eventually defeated the top minds in the subject.

A store of honey

As Benjamin Franklin said, you can catch more flies with honey than with vinegar. But can you store more honey with hexagons or squares?

The honeycomb question is equivalent to something even simpler: suppose I want to divide up a sheet of paper into regions each with an area of $1 \, cm^2$. The most obvious approach, at least to humans, is to use a square grid with each cell $1 \, cm \times 1 \, cm$. But there are other options too. I could divide the page into equilateral triangles (the most symmetric triangles, where all the sides have the same length). Or I could follow the example of the bees, and use a grid that is composed of regular hexagons (six-sided shapes whose sides are all equal).

It is a fundamental fact of geometry that these three shapes, triangles, squares and hexagons, are the only regular polygons that will work. Pentagons or heptagons or octagons, or generally n-gons for higher values of n, cannot be laid side by side without leaving gaps between them. Of these three, which is the most efficient, in the sense of requiring the least total amount of ink, or wax? In mathematical terms, the question is: which has the lowest length-to-area ratio?

Around AD 320, Pappus of Alexandria analyzed these three shapes. He found that a square with area 1 has a perimeter of 4. An equilateral triangle with area 1 has an increased perimeter, of around 4.6. But a hexagon with area 1 has the lowest perimeter, of around 3.7.

It follows that the hexagonal grid requires less ink to draw it than either the square or triangular grid. Lifting this into three dimensions, tubes with a hexagonal cross-section, rather than a triangular or square cross-section, will require the least amount of wax to build of these three possibilities. As Pappus put it, 'Bees, then, know . . . that the hexagon is greater than the square and the triangle and will hold more honey for the same expenditure of material used in constructing the different figures.'

The honeycomb bites back

It was believed that Pappus had completely settled the question of the geometry of the beehive. He had not; he had only compared the three most symmetrical possible designs. But symmetry was not an integral part of the problem. If we return to the problem of dividing a sheet of paper into cells of equal size, there are plenty of other shapes we could use: rectangles, right-angled triangles, or irregular pentagons. In fact, there are endless possible irregular shapes that will work. Indeed there is no reason for the sides to be straight; the walls could be curved, or could even have gaps between them. Could we really be sure that the bees had not missed a better arrangement, which used less wax? This question is hugely more difficult, as the number of possibilities is now infinite. The *honeycomb conjecture* says that the answer remains the same: the plain grid composed of regular hexagons still represents the most efficient method.

Most of the important conjectures in mathematics come with someone's name attached: that of the person who first proposed them. The honeycomb conjecture seems to have slipped into mathematical folklore without anyone's help. Indeed for many years it was even assumed as fact, despite no-one having provided a proof. It was not until 1999 that Thomas Hales finally filled this gap. His theorem was a vindication of the bees: he showed that regular hexagons really do provide the optimal solution.

Humans get the last laugh

It may have taken human geometers nearly two thousand years to prove what bees know instinctively, but in 1953 László Fejes Tóth did manage to get one over on the bees, by finding a way to improve their design. Fejes Tóth's analysis concerned what happens at the end of the hexagonal tubes.

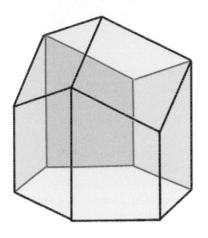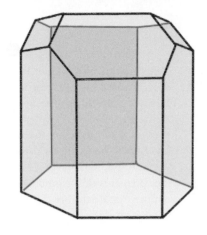

Bees honeycomb (left) and Fejes Tóth's honeycomb (right)

Honeycomb consists of two layers of tubular cells, each open at one end. These two layers are joined back to back. The question is how should the two closed ends fit together? The simplest design would be for both sets of tubes to have flat hexagonal ends. But bees have found a better method. They close off each tube with four rhombuses. (A *rhombus* is a four-sided shape with all the sides the same length, as in a square, but where the edges do not meet at right angles.) This arrangement allows the two layers to mesh together more efficiently.

However, Fejes Tóth was able to identify an even more economical method for closing of the cells than that used by the bees. He showed that sealing each tube with two hexagons and two small squares uses less wax than the bees' rhombuses.

● Kelvin's conjecture

The hexagonal honeycomb conjecture says that an array of hexagons is the most economical way to divide up a page into cells of the same size. In the late 19th century, the mathematician and physicist Lord Kelvin considered the same question in three dimensions. Suppose we have a large room and want to divide it up into small cells, each of the same volume. The simplest design would comprise a lattice of cubic cells, but there are a great many other possibilities. Kelvin's experiments with a variety of shapes led him to conclude that the optimal solution was provided by what is now called a *Kelvin cell*.

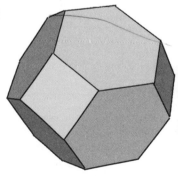

A Kelvin cell

It consists of a shape called a *truncated octahedron*, built from eight hexagons and six squares, with all the faces slightly curved. Kelvin believed that a system built from these cells was the most efficient way to divide up space into equal volumes, minimizing the amount of building material. However, he was not able to prove his idea correct.

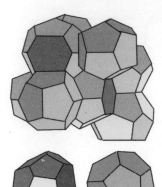

This situation endured until 1993, when two physicists Denis Weaire and Robert Phelan dramatically and unexpectedly refuted Kelvin's conjecture, discovering a new shape that improved on the Kelvin cell. *Weaire–Phelan foam*, as it is known, is a much more complicated shape, but requires slightly less material to construct, 0.3% less in fact. One Weaire–Phelan cell is constructed from six 14-sided shapes, and two 12-sided shapes with irregular curved faces. The resulting foam has a strikingly organic appearance. To commemorate the breakthrough, Weaire–Phelan foam was used in the design of the aquatic centre at the Beijing Olympics in 2008. However, it has still not been established whether this is the best possible solution.

One unit of Weaire–Phelan foam (above) and its two constituent parts (below)

Stacking supermarket shelves: more difficult than expected

Another surprisingly difficult question comes from turning the honeycomb conjecture inside-out. Rather than dividing a page into cells, a different challenge is to squeeze together as many circles as possible.

Suppose I am challenged to fit as many tin cans on a table-top as I can. They are all identical, with round bases. I am not allowed to pile them up, or lay them on their side, just set them upright on the table. What is the best strategy to solve this challenge?

The question amounts to fitting circles together. There are two obvious approaches: the first is to line them up in military rows one behind the other. Each can touches four others, and they all sit on a square grid. The second arrangement uses staggered rows, in which each can touches six others, and they form a hexagonal grid. Out of these two, a little experimentation quickly shows that the hexagonal arrangement is better, covering around 91% of the table, while the squares cover only 79%.

As with the honeycomb conjecture, the real difficulty is to rule out some other, less symmetric arrangement that could squeeze in even more cans. It was in 1910 that Axel Thue proved that this could never be done. The hexagonal arrangement is the best there can ever be.

Kepler's cannonballs

Lifting the circle-packing problem into three dimensions produces another famous problem: sphere packing. If I want to fit as many spheres as I can into a limited volume, what is the best way to pack them together?

How to build the perfect beehive ·

This is a problem solved daily by greengrocers around the world. The traditional way to pile up oranges is first to lay down one layer in the hexagonal packing arrangement, and then lay another similar layer on top, in such a way that the oranges sit as low as they can. Then we put a third similar layer on top of that. In fact there are two choices for the third layer: either it can go directly over the first, or it can be staggered. Each of these two packings fills around 74% of the available space.

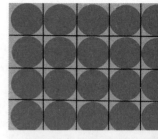
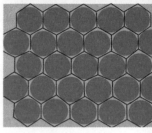

Have the greengrocers missed a trick, or is this really the best possible result? In 1611, the great scientist Johannes Kepler was challenged to find the best way to pile up cannonballs to occupy the least amount of space on board a ship. Kepler came to the conclusion that the greengrocers' method was optimal, but could not completely rule out the possibility of another better method. His claim became known as the *Kepler conjecture*.

This simple observation turned out to be spectacularly difficult to verify. In 1952, Fejes Tóth realized that, to prove it, a number of other possible configurations had to be tested and ruled out. However, that number was astronomically large, and the project was not feasible, at least in the days before the digital computer. In 1998, Thomas Hales managed to put together a proof, which used extensive computer calculations to rule out the various other possibilities.

Two ways of packing circles

Hales' proof was immense, amounting to 250 pages of mathematical argument plus 3 gigabytes of computer calculations. Unfortunately, it defeated even the team of mathematicians assigned to check it. After five years' work, they reported that they were 99% sure the proof was correct, but were unable to fully certify it. Like the four colour theorem (see *How to paint the world in four colours*), Hales' proof of the Kepler conjecture has become an emblem of the change in mathematics that the computer age has brought, in terms of both opportunities and difficulties.

Sphere packing

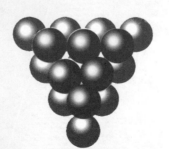

24 How to count to infinity

- The arithmetic of infinity
- Georg Cantor's theory of sets
- Levels of infinity – the countable and the continuum
- The infinite ladder

It has been known for thousands of years that infinity does not behave as ordinary numbers do. David Hilbert provided a famous illustration of this fact. When an ordinary hotel is full, it has no more capacity. If every room is occupied, and a new guest arrives, they must be turned away because there is nowhere to put them. This fact remains true, however large the hotel is. Even if it has a million or a billion rooms, once it is full, it is full.

A weekend at the Hotel Infinity

This sounds so obvious as to be a tautology. Hilbert's *Hotel Infinity*, however, behaves in a strikingly different way. This hotel has infinitely many rooms. Now, suppose that all of them are occupied, and a new guest arrives. This time the concierge can take her bags and welcome her in, as there is plenty of space for her. All she needs to do is have a cup of coffee in the hotel bar, while the hotel manager makes some minor adjustments. He sends a porter to knock on the door of room 1, and politely asks the guest in there to move into room 2. This frees room 1 for the new arrival. At the same time, another porter visits room 2, requesting that the guest there move to room 3. A third porter moves the guest from room 3 into room 4, and so on. In general, the guest from room n is moved into room $n + 1$.

An observer who saw the hotel before and after this manoeuvre, and counted the occupied rooms, would not notice any difference. The number of occupied rooms, and therefore the number of guests, appears exactly the same. What this suggests is that if you start with an infinite collection of objects, and add one more, the result is unchanged. We might be tempted to write this as $\infty + 1 = \infty$. (We have to be circumspect when writing this sort of expression, since the whole point is that arithmetic involving infinity does not behave as ordinary numbers do.)

A similar trick would work even if a hundred new guests arrive. This time, the current guests are asked to move to a room numbered 100 higher than their current one. So the guest in room 1 moves to 101, the guest in 2 moves to 102, and so on, freeing up rooms 1 to 100 for the new guests. So we could write $\infty + 100 = \infty$. It is clear how to extend this method to accommodate any number of new arrivals.

No-one turned away – guaranteed!

In fact the Hotel Infinity even has capacity for infinitely many new guests. Imagine that there is another infinite hotel down the road, the Guesthouse Gargantua, which is also full. Unfortunately, a burst water-main means that it has to close for repairs at short notice. All the guests are asked to vacate their rooms. The question is: can the Hotel Infinity accommodate them all? As we have seen, it can easily take one or a million or any finite number of new guests. But the room-shuffling manoeuvre described above will never make space for infinitely many.

There is a way to do it, however, using a slightly different method. This time, the guest from room 1 is moved to room 2, and the guest from room 2 is moved to room 4. The resident in room 3 is moved to room 6, the one in room 4 is moved to room 8, and so on. In general, the guest in room n is moved to room $2 \times n$. After this process, only the even numbered rooms 2, 4, 6, 8, 10, ... are occupied, leaving all the odd numbered rooms 1, 3, 5, 7, 9, ... available for the exiled guests from the Gargantua. We might tentatively write this as $\infty + \infty = \infty$.

In fact, there is even more space than this. Imagine that the road is infinitely long, and consists of infinitely many infinite hotels, one next to the other. Suppose that they are all full, and then a power cut means that they all have to vacate their rooms, apart from a single hotel. With some even more fancy room-shuffling, all these guests can be accommodated in the Hotel Infinity. This goes to show that $\infty \times \infty = \infty$.

'To see a world in a grain of sand, And a heaven in a wild flower, Hold infinity in the palm of your hand, And eternity in an hour.'

WILLIAM BLAKE, *Auguries of Innocence*

Cantor's powerful sets

The metaphor of the infinite hotel was invented by David Hilbert in the late 19th century. However, the essential properties that it describes had been known for centuries. Infinity is such that when you add or multiply it by itself, any number of times, the result is the same. The moral seems clear: there are no levels of infinity. Whatever you do to it, you can never make it larger. It is a single, undifferentiated monolith.

It was Hilbert's contemporary Georg Cantor who showed that this conclusion is spectacularly and utterly wrong. The hand that delivered this knock-out blow was clothed in a velvet glove. Cantor just needed one simple idea, that of a *set*.

A set is nothing technical, but simply a collection of objects. The set of odd numbers is one example, the set of living human beings is another, and the set of planets in our solar system is a third. As long as there is no ambiguity, there is no problem. So we would have to make clear whether or not the 'odd numbers' include the negative ones, and whether or not dwarf planets such as Eris and Pluto count as 'planets'. As anyone who knows a mathematician will attest, they are very comfortable spelling out exactly what they mean in painfully pedantic terms, so this presents no fundamental difficulty.

What can we do with these sets? Let's start with a simple set, comprising only the numbers 1, 2 and 3, that is, {1, 2, 3} (we always use curly brackets to represent sets). One thing we can do is look at all the *subsets* of this, they are: {1}, {2}, {3}, {1, 2}, {1, 3}, {2, 3} and {1, 2, 3}.

In fact there is one more, called the *empty set*, or Ø. This is a set with nothing in it. To be more precise, it is the only set with nothing in it. It makes sense to class two sets as being the same if they contain exactly the same things (so {1, 2, 3} = {3, 2, 1}, for example). Of course any two empty sets contain exactly the same things, in the trivial sense that they both contain nothing at all. So they must be equal.

Including the empty set, the set {1, 2, 3} has eight subsets: Ø, {1}, {2}, {3}, {1, 2}, {1, 3}, {2, 3}, {1, 2, 3}. Now, there is no reason why we should not have sets of sets. In particular we can look at the set of all subsets of {1, 2, 3}:

$$\{\emptyset, \{1\}, \{2\}, \{3\}, \{1, 2\}, \{1, 3\}, \{2, 3\}, \{1, 2, 3\}\}$$

This is known as the *power set* of the set that we started with: {1, 2, 3}. It is no surprise that this power set contains more elements than the original. In fact, the relationship can be measured precisely. Starting with three elements, the power set contains eight. If we start with a set that contains two elements {1, 2}, then the power set contains four: {Ø, {1}, {2}, {1, 2}}. If we had started with a set of four elements, the power set would contain sixteen (try it!). The rule is that, if the starting set contains n elements, then the power set contains 2^n, that is to say 2 multiplied by itself n times.

● Set theory flexes its muscles

So far, this story about sets and power sets has not been especially exciting. However, when Cantor applied this same simple reasoning to infinite sets, it exploded in spectacular fashion, blowing apart ideas that had stood for thousands of years. In mathematics, there is no shortage of infinite sets. Perhaps the most famous of all is {1, 2, 3, 4, 5, 6, 7, . . .}, which is often known as **N**. What happens when we look at the power set of **N**?

It will include a great many possibilities. Starting with Ø, it will include single-element sets such as {1} and {35}, two-element sets such as {213, 385}, 237-element sets such as {11, 12, 13, . . ., 247}, and a great deal else besides.

There will also be many infinite sets included, such as the collection of primes {2, 3, 5, 7, 11, . . .}, the even numbers {2, 4, 6, 8, 10, . . .}, square numbers {1, 4, 9, 25, 36, . . .} and a host of other random-seeming sets that are almost impossible to describe. When we think about it, this power set seems to contain an awful lot. Cantor showed that this is not an illusion. The power set really is bigger than the original.

The Hotel Infinity: no vacancies

What does it mean to say that the power set of **N** is bigger than **N**? Suppose that the rooms in the Hotel Infinity are labelled by the original set: 1, 2, 3, 4, . . . Now a huge crowd of guests show up, one for every member of the power set of **N**. What Cantor showed is that there is no way that they can all be accommodated. This is true even if the hotel is initially empty. However cleverly you try to reshuffle the rooms, there can only ever be space for a tiny proportion of the guests. Most will be left standing in the car park.

The infinite ladder

The power set of **N** is is known as the *continuum*. The facts that it is bigger than **N** is *Cantor's theorem*. Its proof in 1874 reverberated around the mathematical world. It meant that the concept of infinity is not the indivisible monolith that people had previously thought. It comprises different levels: the infinity of **N**, and then a higher level corresponding to the continuum.

In fact, the full conclusion is even more dramatic. Starting with the set **N**, when we move to the power set, call it P**N**, Cantor's theorem tells us we get something bigger. But then we can repeat the trick, and take the power set of the power set: PP**N**, and this is even bigger. We can keep going, producing a succession of power sets PPP**N** (or P^3**N**), and then $PPPP$**N** (that is P^4**N**), and so on, each infinitely bigger than the last.

'To infinity and beyond!'

Buzz Lightyear, *Toy Story*

With this reasoning, Georg Cantor completely disassembled the concept of infinity that had stood for thousands of years. Far from being a monolith, it is an infinite ladder, with no top rung. However high we have climbed, we can always take a power set and step up to an even higher level. In fact, sets such as $P^{1000}\mathbf{N}$ are the very lowest rungs on the ladder of infinity. To make a bigger step, we can jump up infinitely many steps in one go, and look at the new set this forms: $P^{\mathbf{N}}\mathbf{N}$. We can build ever bigger sets this way: $P^{(P^{\mathbf{N}}\mathbf{N})}\mathbf{N}$.

Such sets may be mind-blowing to most people, but they are small fry for the modern-day set theorist. Today's logicians explore upper reaches of the ladder, totally beyond anything expressed in these terms. Entities here go by names such as 'inaccessible' and 'superhuge', and are defined by their very indescribability, in increasingly powerful languages. Such are the amazing creatures that Georg Cantor released into the mathematical world.

25 How to build a brain

- Complexity and simplicity
- John Conway's Game of Life
- Von Neumann machines and nanotechnology
- Digital physics and artificial life

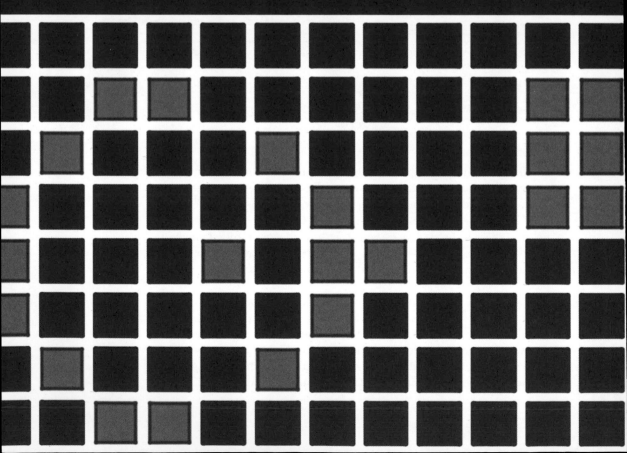

How does complexity arise from simplicity? This question lies at the heart of many areas of science. According to the big-bang theory of cosmology, the entire universe is the result of a tiny speck of matter expanding. The theory of evolution tells us that life on Earth, in all its diversity, derives from simple chemical reactions that took place in the primordial soup 4 billion years ago. On an individual level, each of us has grown from a single cell containing a the blueprint for many details of our body. The works of Shakespeare are built from a collection of a few thousand words, and a few simple grammatical rules. Even the most advanced of modern computer programs are ultimately built from individual instructions for interchanging 0s and 1s within a computer's memory.

These are just some examples of how great complexity can emerge from simple starting points, and this is a hot topic in mathematics too. The mathematics of complexity has major explanatory power in other fields, notably in computer science, and perhaps ultimately in biological sciences and cosmology.

The game of Life

John Conway is one of the greatest mathematicians of the current age. He has made major contributions to a range of subjects including knot theory (see *How to unknot your DNA*), group theory (see *How to slay a mathematical monster*) and game theory (see *How to avoid prison*). Conway also has an interest in recreational mathematics of many kinds and is a talented conjurer in both the traditional and mathematical moulds.

In 1970, Conway invented the game of Life. This 'game' is something of a misnomer, since no human input is required, except in setting up the board at the beginning. With this done, the game then plays itself.

The rules are elegantly simple. The action takes place on a large grid. Each grid-square can be either black ('live') or white ('dead'). Each square then changes colour according to a simple rule, depending on the colours of the eight squares around it. If a square has exactly three living neighbours, then it will be alive at the next moment (either surviving or coming to life). If it has two living neighbours, then it will remain unchanged, living or dead, from its current state. If it has any other number of living neighbours (0 or 1, or 4, 5, 6, 7 or 8) then it will be dead at the next moment (either remaining dead, or dying).

'A mathematician is a conjurer who gives away his secrets.'

JOHN CONWAY

Everything is determined by the starting conditions of the grid, namely the choice of squares that are initially set to living. Once this is decided, and the game left to run, individual squares spring into life, and then die according the rules. The result can be dramatic to watch on a computer simulation, as exotic patterns grow and reverberate around the grid. Sometimes everything eventually dies, apart from a core of stable cells. The most interesting cases are when complex repetitive cycles emerge.

There is an active online community exploring the biodiversity of the game of Life, based at www.conwaylife.com. An entire dictionary of terms has been created to describe and analyze the many patterns that can emerge. There are many millions of possibilities, even starting from small configurations. These can be grouped into families with evocative names such as gliders, guns and spaceships.

Conway invented Life in 1970. Since then the prevalence of the home computer has allowed it to be explored in much greater depth. Computers allow the longer-term trends of the game to be determined, and patterns of particular types to be searched.

The simplest machines

The game of Life is a spectacular illustration of how complexity can emerge from the simplest of rules. With the right initial settings it can even be used to manufacture very large prime numbers. In a sense, the game of Life is not merely an example of complexity arising from simplicity, but it is the archetype.

The device that mathematicians study is called a *cellular automaton*; it resembles the game of Life, but with a few expansions: the number of

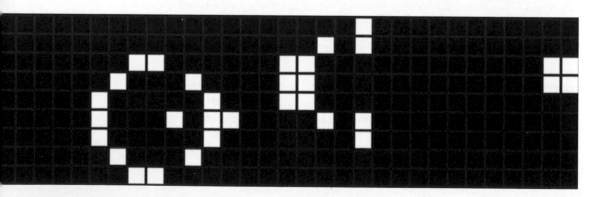

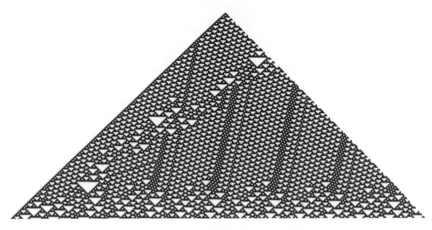

The cellular automaton 'Rule 110'

possible states (or 'colours') of each cell may be more than two, and the number of neighbours of each cell may be more than eight. Finally, the rules according to which a cell changes state may be more complex.

It is known that even simple cellular automata can produce results of the highest possible complexity. In the very simplest automata, the cells lie in a single row, so each cell has two neighbours. Like Life, these very elementary automata have just two colours: black and white. In 2000, Matthew Cook at Wolfram Research established the staggering result that that one such automaton, the 2-colour, 2-neighbour automaton known as 'Rule 110', is capable of *universal computation*. That is to say, it can compute anything that any other computer can. There, literally in black and white, is a wonderful example of the principle of complexity arising from simplicity.

Is the universe a giant computer?

The time and space of the universe seem smooth and continuous, rather than bitty and discrete. Time flows rather than jumps, and space appears as if it can be subdivided into smaller and smaller lengths without ever running into a smallest indivisible length. In the late 1950s, a small group of thinkers including Edward Fredkin and Konrad Zuse pioneered the idea that this is an illusion. Just as when we watch television, or look at a computer screen, we do not notice that the image is formed from individual pixels of different colours, so it is with the time and space of the universe. This philosophy became known as *digital physics*, and it continues to attract followers today.

Is it true? The trouble is that we are not likely to be able to settle this question through direct observation. If the universe is a cellular automaton or something similar, then the individual cells are undetectably tiny. A first guess for the size of a pixel would be the Planck length of around 1.6×10^{-35} metres. This is unimaginably smaller than even the tiniest of subatomic particles, and it is totally beyond our power to see it directly. If the philosophy of digital physics is correct, then this cellular automaton represents the machine-code of the universe, the equivalent to the string of 0s and 1s from which all higher-level computer programs are built (see *How to talk to a computer*). The ultimate aim of digital physics is to find the underlying rules for this automaton, and then prove that the familiar properties of physics can be derived from them: namely quantum mechanics and relativity theory.

The sludge at the end of the world

The game of Life of course exists as a mathematical abstraction, living on a purely notional grid. But with rules so simple, could cellular automata exist as physical devices? If Fredkin's philosophy of digital physics is true, then of course the universe already is such a device, but what if not? In 1948, the mathematician and computer scientist John von Neumann laid out the ideas for a physical machine that could self-replicate. The question is: how simple could such a machine be? The first so-called *von Neumann machines* had only eight different types of component.

Von Neumann did not literally give a blueprint for such a machine, but rather laid out the fundamental properties it would need to have. Since then, the development of *nanotechnology* has brought such a machine a step closer to reality. Eric Drexler was one of the pioneers in this area, investigating the possibilities for machines on the scale of nanometres (one millionth of a millimetre). As well as believing that this technology will bring great benefits, Drexler also outlined the possible dangers should self-replicating *nanobots* ever become reality. Imagine if such machines started replicating out of control, at high speed, incessantly building new copies of themselves from whatever materials they found around them. The idea of an apocalypse through such an explosion of 'grey goo', in Drexler's terms, has subsequently been thoroughly explored in the pages of science-fiction novels (though happily not yet outside them).

Languages and automata

Many linguists study the developments of various human languages in great detail. In 1956, Noam Chomsky took a broader view of the subject. He considered not just languages as they exist in the human world, but as formal structures laid down by abstract rules of grammar. Chomsky ignored all the superficial differences, and classified languages into four strata according to their expressive power: the *regular* languages, the *context-free* languages, the *context-sensitive* languages and the *unrestricted* languages.

How to build a brain

This hierarchy subsequently became important in computer science because each level of language is defined by the complexity of an automaton needed to understand it. The most complex languages, the unrestricted ones, require the most complex form of automaton, equivalent to a Turing machine (see *How to bring down the internet*).

Of course most languages are recognized not by automata, but by human beings. This prompts some fascinating philosophical questions on the relationship between a human mind and an automaton.

Computers come to life

All life on Earth is descended from the very simplest organisms that began replicating in the primordial soup 4 billion years ago. The fundamental rules of life, then, are comparatively simple, just as they are in the game of Life. Yet, over time, the complexity and variety of life, from jellyfish and oak trees to human beings, has emerged.

'All life is an experiment.'

RALPH WALDO EMERSON

In principle, there seems no reason why this process should not take place equally well with virtual entities on a computer. The major factor that biological life has that the game of Life does not is a drive towards complexity over time, through mutation and natural selection. Several systems of artificial life have now been designed, based around cellular automata. *Tierra* was the first, conceived by Thomas Ray in the 1990s. It is a virtual version of the primordial soup, populated by simple self-replicating computer programs (known in other contexts as computer viruses).

To start with, Ray populated the world with just one program written by him and his collaborators. This was the end of the human input to the system, and the program was then left to replicate itself. However, Ray had deliberately built some randomness into the system, so the process of reproduction would occasionally go slightly wrong, resulting in a mutation. Because this happened by chance, the details of the mutation were unpredictable in advance, and it might either be advantageous or not. These various mutations could then compete for resources (in this case computer memory and processor time). Those that had what they needed would survive and reproduce, and the rest would die off.

A sophisticated virtual ecology emerged, with parasitism being an early phenomenon. This is an evolutionarily important observation, since the cohabitation of simple organisms is how complex ones, such as you and me, eventually form.

26 How to bring down the internet

- Computer programs and their different speeds
- The hare and the tortoise: how to tell the difference
- Integer factorisation and internet security
- The puzzle of P versus NP

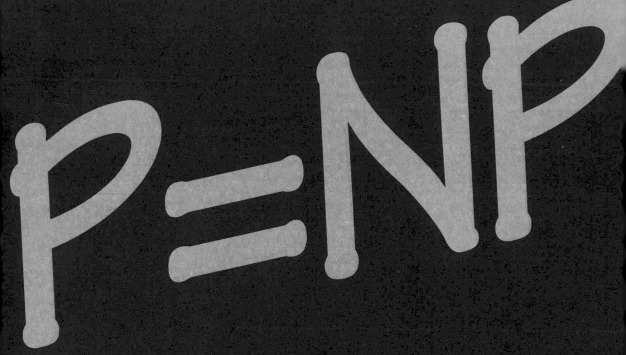

Academic journals and international conferences are the traditional ways that mathematical breakthroughs become public knowledge. One of the biggest open questions in mathematics today may announce itself through a completely different channel. It is the P = NP problem. If this is solved, the first we may know about it is when strange things start happening around the internet: money mysteriously appearing and disappearing, the toughest security systems being waltzed through as if they were not there.

The computer's family-tree

It is impossible to pin down the exact moment that the computer was invented. Was it when Gottfried Leibniz designed his *stepped reckoner* in 1694? Or was it Charles Babbage's *difference engine* of 1847? These are some of the physical forebears of the modern computer. Perhaps the crucial theoretical breakthrough, however, was Alan Turing's development of the *Turing machine* in 1937. Despite its name, this was a purely notional device, rather than a physical appliance such as Leibniz's or Babbage's. It was not until much later that real machines emerged, able to carry out Turing's vision. Nevertheless his insights had a seismic impact even before this.

A Turing machine is a device for carrying out lists of commands, with complete accuracy. Although the individual commands have to be of a very simple kind, the advantage of a machine over a human is that they never make mistakes or get bored. A machine will keep crunching away at its calculation, even if it takes many years to complete. This allows simple commands to be combined into larger packages, which tell the machine how to perform calculations, even of extremely long and complex forms beyond the power of a human alone. Mathematicians call these lists of instructions *algorithms*. Most other people know them as *computer programs*. With this critical idea, the repertoire of the computer was extended far beyond the number crunching of the early calculating machines, and the scene was set for the world of graphics and gaming that we know today.

The billion-year computer program

One of the critical questions concerning algorithms is how long they take to complete. A program that takes millions of years to do its job is of no practical use. This is the basis on which internet security operates. There is no obstacle, in principle, to someone cracking the code that safeguards your bank account. Even the most amateur of hackers could write a program that would do the job . . . eventually. Essentially it amounts to cracking a combination lock with a hundred-digit code. It's easy: first try 000 . . . 000, then 000 . . . 001, and then 000 . . . 002, and carry on until you find the right number. There is nothing inherently complicated about this, the only trouble is that it could take up to 999 . . . 999 steps to complete. Very likely no humans will be around to see the task finished.

This goes to show that there is an important divide between algorithms that run quickly enough to be of practical value, and those that trundle on for millennia. There is a theoretical basis to this divide, which is the subject of *complexity theory*, occupying a niche on the border between mathematics and computer science.

● The hare and the tortoise: how to tell the difference

'*Machines take me by surprise with great frequency.*'

Alan Turing

Suppose I write a simple computer program for multiplying numbers together. It might be that on three pieces of data (that is to say, three digits), my algorithm takes 9 steps to arrive at the answer. On four pieces of data it takes 16 steps, on five it needs 25, and so on. The general rule is that on n pieces of data it will take $n \times n$ steps, that is to say n^2. This is reasonably fast and therefore potentially a useful algorithm. Even on 100 pieces of data, the number of steps required is only 10,000: less than a second's work for a modern computer.

In contrast to this, the combination code algorithm described above takes 10 steps on one piece of data (that is to say, if the code is one digit long), 100 steps on two, 1,000 on three, and so on. In general, for n pieces of data, it will need 10 multiplied by itself n times, that is 10^n, steps to complete. This quickly grows out of control. The hundred-digit code amounts to 100 pieces of data, by which point the number of steps needed is 10^{100}, far exceeding the number of atoms in the universe. So this algorithm is essentially useless for most practical purposes, and will remain so no matter how fast computer processors become.

The difference between these two algorithms can be seen in the two algebraic expressions: n^2 and 10^n. The first of these is what mathematicians call a *polynomial*, while the second is an *exponential*.

● Do not exceed n^2 miles per hour

Today's computer programmers are experts in finding efficient algorithms for carrying out tasks, but there are limits to what even the most accomplished programmer can achieve. Every task comes with its own

How to bring down the internet

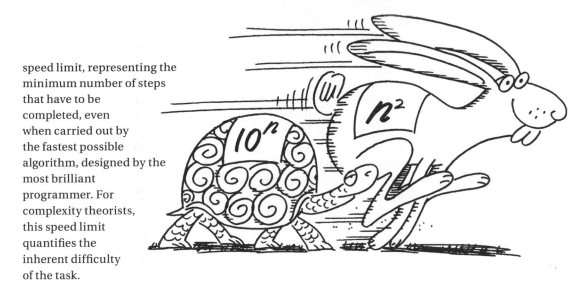

speed limit, representing the minimum number of steps that have to be completed, even when carried out by the fastest possible algorithm, designed by the most brilliant programmer. For complexity theorists, this speed limit quantifies the inherent difficulty of the task.

Tasks whose limit is given by a polynomial, such as n^2 or n^3, are said to be 'in P'. (P is the collection of all such tasks, standing for 'Polynomial Time'.)

The significance of the class P was pointed out by Alan Cobham. He said that tasks that are in P are essentially those that are tractable in the real world. They can practically be carried out within sensible amounts of time, on a real computer. Tasks that are not in P generally cannot. There may be algorithms that can compute them of course, but they might take centuries to complete.

Cobham's rule is something of a generalization, but it is valuable as a rule of thumb. However, it prompts a seriously difficult puzzle: how can we know whether or not a given task is in P? Well, if we write an algorithm to carry out our task quickly, then we know that it is in P. That is straightforward enough, but the reverse is altogether tougher. How can we ever be sure that a task is not in P? How can we ever know when we have found the fastest possible algorithm to complete a given task? This question defines the limits of what a computer can do.

● The internet's skeleton key

It is one of the most fundamental facts of mathematics that every whole number is built from prime numbers (see *How to win the ultimate maths prize*). For example, $6 = 2 \times 3$ and $66 = 2 \times 3 \times 11$

The question is: if we are just presented with a number such as 924, how can we find what its prime components are? This question is known as the *factorization problem*.

In principle, it is easy. We just try out each prime in turn to see how many times it divides into 924. We start by trying to divide by the smallest prime, 2. We can do this, with a result of 462. Then we can divide by 2 again, to get 231. Now we cannot divide by 2 any more, so we move on to the next prime, 3. This does divide 231, giving a result of 77. We cannot divide by 3 any more, so we move on to the next prime 5. This doesn't divide at all, so we move straight on to 7. This does divide it, giving 11. Since this is a prime number, we have finished, and the answer is $924 = 2 \times 2 \times 3 \times 7 \times 11$.

This method will always work, no matter what the starting number. Apart from being rather tedious, the trouble is that for large numbers it becomes inordinately slow. Modern internet encryption relies on numbers hundreds or thousands of digits long. To factorize such a number using this method would require an outrageous number of steps, totally beyond the capacity of any computer.

This raises the central question of modern internet security: is there a quicker method for factorizing numbers? Is there some clever mathematical trickery we might exploit to factorize large numbers speedily? In terms of complexity theory, the crucial question is: is the integer factorization problem in P? If it is, and a fast method could be found, it would act as a universal pass-key to even the most secure of internet sites.

This may sound implausible, but mathematics is full of surprises. In 2002, Manindra Agrawal, Neeraj Kayal and Nitin Saxena published a breakthrough paper entitled 'Primes is in P'. In this, they showed that a closely related task can be solved quickly. Namely, if you start with a large number, and want to know whether or not it is prime, there is a fast algorithm that can do the job. This was an unexpected development.

Slow to do, fast to check

Although the integer factorization problem is slow to carry out, it is very quick to check the answer. If I am asked to factorize 2,491, it will take me some time. On the other hand, if someone asks me to verify that the answer is 47×53, being fairly adept at long multiplication, I can do it quickly (even faster with a calculator). This suggests that a new class of tasks to consider are those that can be checked quickly, even if they cannot initially be performed quickly. The jargon for such problems is that they are 'in NP'. (NP stands for 'non-deterministic polynomial time', meaning that the time taken to check the answer is polynomial.)

We now have two classes of task: firstly, those that can be carried out quickly. Multiplying two numbers is one such problem. This collection of all such problems is known as *P*. The other class consists of those tasks that can be checked quickly. Integer factorization is one. This collection of problems is known as *NP*.

How do these two classes relate to each other? It is certainly true that every problem that can be solved quickly can also be checked quickly, so *P* is a subset of *NP*. The more difficult question is the other way around. Is it true that every problem that can be checked quickly can also be solved quickly?

The obvious answer is 'no'; there seems no sensible reason to suppose that this should be true. Indeed common sense seems to militate against it. Amazingly, however, mathematicians have as yet failed to find a single example of a problem that can be checked quickly, which we also know for sure cannot be solved quickly.

There is no shortage of contenders for such a problem. Integer factorization is one; the travelling salesman problem is another (see *How to visit a hundred cities in one day*); the unknotting problem a third (see *How to unknot your DNA*). Although many people suspect that these problems can never be solved quickly, no-one has managed to prove this for certain. This is the old difficulty of proving a negative.

It therefore remains possible that what we least expect will turn out to be true: that these and many other problems will in fact turn out to have quick solutions. In other words, that $P = NP$. If true, this would be one of the most astonishing and dramatic twists in the whole history of mathematics and computation.

Even if this is not true, as many suspect, knowing for certain that $P \neq NP$ would herald a new era in understanding of the theoretical limits of algorithms. This question of whether or not $P = NP$ is one of the Clay Mathematics Institute's Millennium problems, and so now comes with a $1,000,000 price tag for anyone who manages to resolve it one way or the other. One thing is for certain: $1,000,000 is a wild underestimate of the true value of this problem.

27 How to ask an unanswerable question

- Eubulides' Liar paradox
- Gottlob Frege's foundations of arithmetic
- Bertrand Russell's paradox
- *Principia Mathematica*

The most ancient and purest of all paradoxes, The Liar, was first conceived by the philosopher Eubulides in the fourth century BC. *With the few short words 'This is a lie', he demolished an entire world-view: that every statement must fit into one of two boxes, labelled 'true' and 'false'.*

Eubulides' Liar refuses any such classification: if the statement is true, then it must be false. On the other hand, if it is false, then the speaker is telling the truth after all. The Liar dances between the two boxes, refusing to settle in either.

At the root of the paradox is self-reference. The Liar exploits the ability of a sentence in English (or originally Greek) to describe itself. There are various ways to dress up the Liar paradox in fancier clothes. For example:

1. Statement 2 is true.
2. Statement 3 is true.
3. Statement 1 is false.

Such paradoxes are delightful mind-bending conundrums, but do they have any practical importance? The answer is emphatically yes. Thousands of years later, Eubulides' Liar crept unseen into the world of mathematics. Twice during the 20th century, it came dangerously close to bringing the whole edifice crashing to the ground. The steps that mathematicians took to avoid this disaster went on to transform the subject.

> *'This is a lie.'*
>
> EUBULIDES

Sets and ladders

In the late 19th century, Georg Cantor shocked his contemporaries by disassembling the concept of infinity. He showed that, rather than infinity being a single entity, there is an infinite ladder of infinities, with each rung infinitely higher than the last (see *How to count to infinity*). Love it or hate it, no-one could deny that Cantor's idea was truly spectacular. It was all the more astonishing that it hinged upon a single, very simple idea.

A *set* is nothing more than a collection of objects. This straightforward definition seemed to form a solid basis for Cantor's investigations. At least it appeared that way, until the logician Bertrand Russell found Eubulides' ancient paradox lurking at the heart of the fashionable new theory.

Russell's paradox

If a set is just a collection of things, then there is no reason why we should not have a set of sets. In fact, we might even talk about the set of all sets. Now, this has a very unusual property: it includes itself as a member.

Bertrand Russell sensed that self-including sets could give rise to the same paradoxical predicaments as sentences that describe themselves. Most familiar collections of objects, of course, do not behave in this strange way. So Russell carved the universe of sets into two: the *exotic* sets, which include themselves, and the *plain* sets, which do not. Russell's paradox concerns the collection of all plain sets. Call this set R. Russell's unanswerable question was: is R exotic or plain?

To see the problem, assume that R is exotic. That means that it includes itself. But the only sets that qualify for membership of R are the plain ones, so this is impossible. On the other hand, suppose R is plain. Then it satisfies the membership criterion for R, and so is included in R. That means R is exotic. Just as for with the Liar paradox, either assumption leads to a contradiction.

Russell also came up with an analogy to illustrate the idea in the real world, sometimes known as the *paradox of the barber*. In a village, all the men are shaved every day. Some shave themselves, and the rest are shaved by the town's barber. In fact the barber shaves exactly those men who do *not* shave themselves. The question is: does the barber shave himself or not?

Russell realized the far-reaching importance of his paradox, and knew that it was not just Cantor's infinities that it endangered.

● Frege's foundations

Ever since the time of Plato, mathematicians and philosophers have debated the nature of mathematical objects such as numbers. To what extent can numbers really be said to *exist?* Are they only in the mind, or do they have some objective reality?

An easier question than what numbers are is what they are *for.* Since the dawn of history, numbers have been used primarily for one purpose: counting. To put it another way, numbers measure sets. This was the basis on which Gottlob Frege had attempted to provide the first rigorous underpinning of mathematics.

What is the number 2? This is a rather difficult question. It is much easier to identify instances of it: two bananas, two cars, two people, and so on. Frege's idea was to define the number 2 as the complete collection of all such pairings. Then, using only basic properties of sets and laws of pure logic, Frege was able to build up all the familiar arithmetical properties of numbers.

Frege's great work, *Grundgesetze der Arithmetik* (*The Basic Laws of Arithmetic*) was shaping up to be a magnificent achievement. It was a cruel blow when Russell intervened with his paradox, just as the work was going to print. The revelation that sets are inherently contradictory was devastating to Frege's project.

Principia Mathematica

Frege's careful work cruelly unravelled before his eyes. But his aim, to put mathematics on a firm logical footing, had been a noble one, which was subsequently taken up by Russell himself. Teaming up with his former teacher Alfred North Whitehead, Russell set to work on a three-volume masterpiece, *Principia Mathematica*.

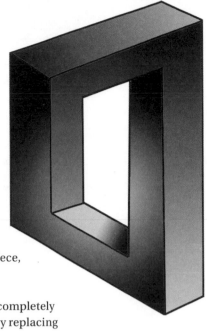

A visual paradox

Principia Mathematica was a technical tour de force, and almost completely unreadable. It managed to resurrect many of Frege's arguments by replacing sets with more complicated and less intuitive objects called *types*. Unlike sets, every type comes with a level. The level 0 types resemble ordinary sets of objects, the level 1 types are sets of sets, the level 2 types are sets of sets of sets, and so on. Generally, types are permitted only to include other types of lower level. Crucially then, no type can ever include itself.

This theory of types introduced a new layer of technicality on top of Frege's already abstruse arguments. Amazingly, it took until page 83 of the second volume of *Principia* to deduce that $1 + 1 = 2$ (at which stage the authors drily observe 'The above proposition is occasionally useful'). The pay-off for this stupendous effort was that Russell's contradictory monstrosity was avoided. The laws of arithmetic, along with Cantor's theories of infinity, had survived the first modern attack of the paradox.

Axioms and proofs

In *Principia Mathematica*, Russell and Whitehead had achieved something truly majestic. But what exactly was it? The target of their work was the very heart of mathematics.

Anyone can put their feet up and lazily opine about mathematics: 'no, there are no even perfect numbers', 'yes, the Riemann hypothesis is true'. What distinguishes a *theorem* from a mere guess or conjecture is that someone has provided a watertight argument in support of it. This is *proof*, the magical ingredient that separates mathematics from every other field of thought.

More precisely, a proof is a sequence of logical deductions culminating in the theorem. Constructing a proof is painstaking work, as each stage in the argument needs to be so simple as to be obviously true. For a result as complex as the Poincaré conjecture, for example, an immense number of such steps may be required.

Proofs cannot be conjured from thin air, however. At the beginning of every proof must be some starting assumptions, called *axioms*. This seems a reasonable requirement, but it prompts a very difficult question indeed: if mathematicians are in the business of investigating the behaviour of the ordinary numbers 0, 1, 2, 3, 4, 5, . . ., then what should these underlying axioms be? What are the fundamental axioms of arithmetic?

● Hilbert's dream

The investigations into the axioms of arithmetic had begun in earnest during the 19th century, in the work of Giuseppe Peano, Gottlob Frege, and others. The insights of these two men were both incorporated into the monumental *Principia Mathematica*.

This intellectual snowball effect was certainly picking up speed, but where was it heading? The ultimate goal was articulated by David Hilbert in the early 20th century. What he hoped for was a system to provide a logically watertight basis on which arithmetic could be built. It should comprise some underlying axioms, as well as a set of rules for proof that specify exactly which deductive steps are legitimate and which are not.

Hilbert set out some criteria that such a system should satisfy. Most importantly, it should be *consistent*. That is to say, following the rules of the system should never result in a nonsensical outcome such as $2 + 2 = 5$.

The second requirement was more ambitious: the system should be *complete*. This would mean that it captured the entirety of arithmetic, that every true statement about numbers should be deducible within the system. So there must be no gaps: for any assertion (call it H), either H or its opposite 'not H' should be deducible within the system.

'How wonderful that we have met with a paradox. Now we have some hope of making progress.'

Niels Bohr

How to ask an unanswerable question

With the discovery of a system of this kind, Hilbert believed, mathematicians could truly claim to have tamed the wild world of numbers. It was a grand and inspiring goal. With the arrival of the *Principia Mathematica*, and with Russell's paradox out of the picture, the dream seemed tantalizingly close. But it was not to be. The liar paradox was to make another dramatic entrance, and this time no amount of hard work would be enough to avoid it.

Gödel's incompleteness theorem

Principia Mathematica attracted interest around the world. But, in Vienna, a young logician by the name of Kurt Gödel devoted a great deal of time to studying the system that Russell and Whitehead had carefully built, inspecting it for cracks.

The key to the liar paradox had been the capacity of the English language to refer to itself. Gödel realized that the language of the *Principia* was complex enough for the same trick to work. In 1931, he dealt Hilbert's hopes a devastating blow. By inventing an ingenious way to encode proofs within arithmetical statements, Gödel was able to translate the following statement into the language of the *Principia*: 'this statement has no proof'.

It was a paradox exactly in the mould of Eubulides'. If we call Gödel's statement G, then G can either be proved or it cannot. If it can, this is a disaster: it means that the rules allow a false statement to be derived, and the system is inconsistent. The alternative is that G cannot be proved. This is harmless on one level: G is true, and the system remains consistent. However, G now forms an example of a statement that is true, but that cannot be proved within the system. Therefore the system is incomplete.

Gödel's theorem did not simply mean that *Principia Mathematica* had failed in its goal of encapsulating all of mathematics. It applied with equal force to any possible system that any future mathematician could come up with. Hilbert's dream of a complete, consistent system underlying arithmetic was wholly unrealizable. Nothing mathematicians could write down (or later program into a computer) would ever be able to capture all the richness of arithmetic.

The great mathematician John Von Neumann was at the lecture where Gödel announced his historic result. At the end, he was heard to remark 'it is all over'. In fact, as every computer-owner knows, the age of logic was only just beginning.

28 How to detect fraud

- Simon Newcomb's first digit phenomenon
- Frank Benford's law
- Scaling and rescaling
- Theodore Hill's theorem

The modern world is full of statistics. From football leagues and lottery results to global temperatures and city populations, almost everywhere humans look we see situations to describe and analyze with numbers. These numbers can have many different sources, some arising from human activity whether in sports or in business, and others describing purely natural phenomena such as geographic data. Others are essentially random. Nevertheless there are some broad features that many sets of data have in common. In the late 19th century, the astronomer Simon Newcomb noticed one highly unexpected pattern recurring in many collections of data.

Take any set of numbers, such as the areas of river basins worldwide. Newcomb's observation involved the *leading digit* of each number, meaning the first digit to appear. So, for example, 7,058 has leading digit 7. We ignore any zeroes at the beginning, so 0.00501 is deemed to have leading digit 5.

The question is: how often would you expect each of the digits 1, 2, 3, 4, 5, 6, 7, 8 and 9 to appear as the leading digits? The obvious answer is that, on average, each should appear equally often. Of course there may be a bit of variety, but this would be the general rule; at least this was what most people expected. However, when Newcomb actually performed this experiment, he was astonished to find something entirely different. Using data from a range of different sources, a general trend became clear: not all leading digits are equally common.

Some numbers are commoner than others

In the years before pocket calculators, the *log table* was a source of torment for generations of reluctant maths students. 'Log' is short for *logarithms*, useful devices for multiplying and dividing large numbers.

Simon Newcomb noticed that, in the rather old and tatty book of log tables he was using, the earlier pages were significantly more thumbed than the later ones. It is a tribute to his scientific curiosity that rather than dismissing this chance observation, Newcomb set out to explain it. Now, a book of log tables is not like a novel: you do not begin at the beginning, and read through to the end, not unless you wish to die of boredom. Rather, it is like a dictionary: you look up what you need. If the earlier pages had seen more usage than the later ones, this suggested that people performed calculations with the numbers earlier in the book more often than they did with those later in the book.

Newcomb began examining sets of data in the real world, and looking at the propensity of each digit to appear first. He found that the digit 1 occurs as the leading digit much more commonly than all the others. In fact, 1 is the leading digit around 30% of the time. The remaining digits occur with decreasing frequency, according to the graph below.

Where did this pattern come from? Was it a one off, perhaps a curious by-product of the particular type of data that Newcomb was looking at? No-one had an answer until 1938, when the physicist Frank Benford followed in Newcomb's footsteps. He made the same chance observation about log-table usage, and then rediscovered the same strange pattern as Newcomb. Benford then set out to test this theory in a very wide range of scenarios. He analyzed baseball statistics, the areas of river basins, death rates, the square roots of various numbers, the addresses of American mathematicians, the weights of molecules, and miscellaneous numbers appearing throughout a magazine, among many other sources of data. In almost every case, he found that the distribution of leading digits closely followed the rule that he and Newcomb had discovered.

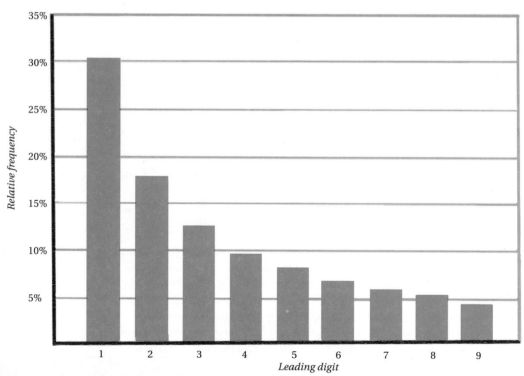

How to detect fraud

Newcomb and Benford were each able to find a formula to capture this pattern. This formula later became known as *Benford's law* (somewhat unfairly to Newcomb, who found it first). In technical terms, it says that the digit n should occur as a first digit a proportion of the time given by $\log_{10}\left(1 + \frac{1}{n}\right)$. Benford's law turned the mysterious first-digit phenomenon into something approaching a mathematical conjecture. But it still gave no hint whatsoever about why this bizarre phenomenon should hold true.

Between too random and not random enough

Benford's law has been observed in a wide variety of contexts, but it is not always valid. If you perform an experiment picking numbers at random between 1 and 99, the results will be *uniform*: on average each digit will occur equally often as the leading digit. Another case where the law fails is when numbers come within a very constricted range. If we measure the ages (in years) of children at a kindergarten, we cannot expect Benford's law to apply: they will largely have leading digits between three and six. Similarly if we measure the length in hours of films at the cinema or the heights in metres of horses in a stable, these will all be concentrated within a narrow range, drastically limiting the spread of leading digits.

Between these two scenarios, however, are innumerable situations where Benford's law does apply, with astonishing accuracy. Ever since his work, the same question has been asked over and over again: why on Earth should Benford's law be true?

Scaling and rescaling

If we measure the heights of a mountain range, say, we would not expect the distribution of first digits to be seriously affected by the choice of units. If we measure them in metres, or in inches, or in feet, every individual number will be changed, but the overall pattern should remain roughly the same. This amounts to saying that the distribution of first digits should be *scale invariant*; it should remain the same under rescaling.

If there is a universal law of leading digits, it must satisfy this rule of rescaling. Rescaling, or changing the units of measurement, amounts to multiplying by some fixed number. In the case of converting feet to yards, this number is 3. When we apply this line of thought to uniform distributed data, where all digits are equally prevalent, it fails to satisfy this requirement. For example, if we start with a completely uniform set of data:

1, 1.5, 2, 2.5, 3, 3.5, 4, 4.5, 5, 5.5, 6, 6.5, 7, 7.5, 8, 8.5, 9, 9.5

Here each first digit does indeed occur equally often. When we change the units by multiplying by 2, we get some new numbers:

2, 3, 4, 5, 6, 7, 8, 9, 10, 11, 12, 13, 14, 15, 16, 17, 18, 19

Here the first digits are heavily skewed towards 1. If we instead multiply by 3, we get:

3, 4.5, 6, 7.5, 9, 10.5, 12, 13.5, 15, 16.5, 18, 19.5, 21, 22.5, 24, 25.5, 27, 28.5

Again, the leading digits 1 and 2 are much more highly represented than the others. This illustrates that the uniform distribution on first digits is not scale invariant.

However, the distribution given by Benford's law is scale invariant. If we start with a set of data that obeys Benford's law, and then rescale the whole lot, the rescaled data will also obey Benford's law. This was the first convincing partial explanation for Benford's law. The second was even more persuasive: Benford's law represents the *only* scale invariant distribution of first digits. No other distribution of first digits has this property of scale invariance.

Benford goes base-jumping

Benford's law, as stated above, only relates to numbers written in base 10, that is, in our ordinary decimal notation for numbers (see *How to talk to a computer* for more discussion on different bases). This is how we traditionally write our numbers, but it has no mathematical significance. So if there is to be a universal law of first digits, then it should account for the first digits in every base, not just 10. In 1995, Theodore Hill showed that Benford's law is the only possible distribution of leading digits that satisfies this notion of base independence. If we start with a set of data in base 10 that satisfies Benford's law, and then convert it to base 3 say, the result will generally satisfy the equivalent condition. Technically this says that the digit n, when working in base b, should occur as the leading digit a proportion of the time given by $\log_b\left(1 + \frac{1}{n}\right)$.

Hill mixes it up

The arguments from scale invariance and then base invariance were fairly persuasive accounts of Benford's law, as far as they went. But they did not really come to terms with the central problem: many sets of data obey Benford's law, but some do not. Benford himself, when looking at his data, noticed that not every source of data produced a perfect match with his law.

The idea of data being random enough, but not too restricted, is correct, but it is hardly mathematically rigorous.

The key to this was noted by Benford himself: the best fit of all came when he looked at the data across his whole table. When he jumbled up atomic weights together with street addresses, physical constants, stock-market data, the resulting mixture satisfied his law almost perfectly.

This was suggestive: data in the world comes from many different distributions, the uniform distribution of lottery numbers, the normal distribution of human heights (see *How to have beautiful children*), and so on. When these are all mixed together to produce a 'random distribution', and the first digits of the result taken, Benford's law emerges. This was the idea, but to make it rigorous required some technically advanced probability theory. This was accomplished by Hill in 1995, completing the explanation of Benford's law.

Benford goes to court

Benford's law is certainly an unexpected quirk of the place-value system of writing numbers. Its sheer unexpectedness makes it a very useful fact, not in mathematics particularly, but in the wider world of statistics. One useful application is in the detection of fraud, pioneered in the 1990s by Mark Nigrini. When crooked businesspeople cook their books, it seems that they are usually unaware of Benford's law. On the other hand, research suggests that legitimate accounting data does obey it. To an investigator aware of Benford's law, the falsified data, which usually has first digits that are approximately uniformly distributed, will stand out like a sore thumb.

As discussed above, there are situations where Benford's law does not apply, so the failure of Benford's law on its own is no guarantee of criminality. Nevertheless it can provide strong supporting evidence, and has successfully been used to help convict several fraudsters in the USA.

'After all, facts are facts, and although we may quote one to another with a chuckle the words of the Wise Statesman, 'Lies – damn lies – and statistics', still there are some easy figures the simplest must understand, and the astutest cannot wriggle out of.'

LORD COURTNEY (1895)

29 How to create an unbreakable code

- Secret writing: cryptography and cryptanalysis
- Monoalphabetic and polyalphabetic codes
- Uncrackable codes: the one time pad
- Keeping your money safe: public key cryptography

LeZTJUKTVeQULLStteEMBSZRQUZMSReVI
JZDMJUTTUOMteFTMTDVTMXeUXTeUJKT
DZeteDVtKOeQULeSZXeDQeOeKDeEteQKZU
ITUTVDReDZRSZJUVlDtSUZtUSLXDVTXTe
RESMDYteIUHVOeDXUZMMIMTeLMDRZA
THMDQQeMMTUIUHVMeQHVSTIZeTOUV
eFteZMRStKSZBtKeSRSUTMLASKTJDTTJUV
SAKtSKDeeQTeDVDZQetUXVUQeEROSTKS
RSTDeTVDZMLSMMSUZReDVAeZeVDTIDI
LeZTJUKTVeQULLStteEMBSZRQUZMSReVI
JZDMJUTTUOMteFTMTDVTMXeUXTeUJKT
DZeteDVtKOeQULeSZXeDQeOeKDeEteQKZU
ITUTVDReDZRSZJUVlDtSUZtUSLXDVTXTe
RESMDYteIUHVOeDXUZMMIMTeLMDRZA
THMDQQeMMTUIUHVMeQHVSTIZeTOUV

Over the course of humanity's bloody history, there has been a near constant need to communicate securely with your own side's spies, troops and allies, while preventing the enemy from listening in. This is the science of cryptography. At the same time we want to intercept our enemy's communication and discover their secret plans. To do this we need to break their ciphers. This is the art of cryptanalysis.

Breaking an enemy's code can prove a decisive moment in war-time. The most famous recent example was during the Second World War, when the allies cracked both the Enigma code in Europe and the Purple code in the Pacific, gaining key strategic advantage. One of the many people involved was the great mathematician Alan Turing (see *How to bring down the internet*).

The science of secrecy

One ancient form of secret writing is to change the individual letters of the message according to some rule. This is a *cipher*. The simplest example is when the alphabet is jumbled up before the message is written. This is called a *monoalphabetic cipher*. One possible key is as follows. (It is customary to use lower case letters for the plain text, and upper case letters for the enciphered text.)

a	b	c	d	e	f	g	h	i	j	k	l	m	n	o	p	q	r	s	t	u	v	w	x	y	z
Z	V	M	D	H	L	E	Y	O	X	B	A	F	J	Q	R	U	P	C	N	S	G	K	W	T	I

To communicate, both the sender and receiver must have access to this. Then messages can easily be encrypted and decrypted. To make the message harder for any interceptor to decipher, punctuation and spaces are omitted:

NYHHZPNYAOJECZPHVHMQFOJECSCROMOQSCYQADTQSPRQCONOQJ
CZJDDQJQNPHGH ZATQSPODHJNONOHCKHKOAAPHZNNHFRNMQJNZM
NJHWNMHJNSPT

Al Kindi's code-cracking

Suppose you intercept an enciphered message. If you do not have the key, how might you try to break the code? In ninth-century Iraq, the scholar Al Kindi made a critical observation while studying the Koran: not all letters in Arabic occur equally often. The same applies in English of course, and indeed in all alphabetic languages.

From this harmless-seeming observation, Al Kindi developed a powerful technique called *frequency analysis*. It works by counting the commonest symbols in the enciphered message, and then attempting to replace these with the commonest letters in English. So we have to know which these are. The illustrated table of the English alphabet gives the average frequency of each letter.

Letter	Frequency	Letter	Frequency
e	12.7%	m	2.4%
t	9.1%	w	2.4%
a	8.2%	f	2.2%
o	7.5%	g	2.0%
i	7.0%	y	2.0%
n	6.7%	p	1.9%
s	6.3%	b	1.5%
h	6.1%	v	1.0%
r	6.0%	k	0.8%
d	4.3%	j	0.2%
l	4.0%	x	0.2%
u	2.8%	q	0.1%
c	2.8%	z	0.1%

Suppose we now intercept a message intended for an enemy:

YGHDHYOQDHRQHOWJSOQXXQITHYDHPXOQWWQMGPCDXAGFCOOQ
XXGGWBQIYFCIWQYGDHXQCIXGKXWXHDXWEGCERGCPXAGERHIGXG
HDXANGFCOGQIEGHFGNGSDQIT XGFAICRCTZXCXDHYGHIYQIPCDO
HXQCIXCQOEHDXYQWH SRGZCJDNGHECIWZWXGOWHIYHRRCNJW
QIXCZCJDWGFJD QXZIGXNCDBWXGKXGIYW

Without the key, this presents a challenge. However, if we assume this was encrypted using a monoalphabetic cipher, we can try to apply frequency analysis. The first thing to do is to analyze how often different letters appear:

G	X	H	C	Q	I	W	D	Y	O	R	E	F	J	N	Z	A	P	S	T	B	K	M	L	U	V
26	26	19	19	18	16	15	15	10	10	7	6	6	5	5	5	4	4	3	3	2	2	1	0	0	0

How to create an unbreakable code

Since the commonest two letters in the cipher text are G and X, we might guess that these represent the commonest two plain text letters, e and t respectively. If this is right, we have partially deciphered the text:

YeHDHYOQDHRQHOWJSOQttQITHYDHPtOQWWQMePCDtAeFCOO
QtteeWBQIYFCIWQYeDHtQCIteKtWtHDtWEeCEReCPtAeERHIeteHDtA
NeFCOeQIEeHFeNeSDQITt eFAICRCTZtCtDHYeHIYQIPCDOHtQCItC
QOEHDtYQWHSReZCJDNeHECIWZWteOWHIYHRRCNJWQItCZCJDWeFJ
DQtZIetNCDBWteKteIYW

It is not just individual letters that are used in frequency analysis. Some pairs of letters are much more common than others, the commonest of all being 'th'. In the above message, the decrypted letter t is followed on three occasions by the encrypted letter A. This suggests that A may represent the letter h. Exploring this possibility, we get:

YeHDHYOQDHRQHOWJSOQttQITHYDHPtOQWWQMePCD theFCOO
QtteeWBQIYFCIWQYeDHtQCIteKtWtHDtWEeCEReCP theERHIeteHDth
NeFCOeQIEeHFeNeSDQITteFhICRCTZtCtDH YeHIYQIPCDOHtQCItCQ
OEHDtYQWHSReZCJDNeHECIWZWte OWHIYHRRCNJWQItCZCJD
WeFJDQtZIetNCDBWteKteIYW

By continuing in this vein, trying to replace the commonest encrypted letters with the most frequent letters in English, it should be possible to completely decipher the message. However, frequency analysis is something of an art. One cannot mindlessly substitute letters according to frequency and expect to crack the code. 'cz' may not be a common combination of letters in general, but if the message relates to the Czech republic, then that combination may appear often in a particular message. It is often necessary to try out different possibilities. Usually a much lengthier piece of text is required; this example is quite artificial.

Code-makers and code-breakers: an arms race

Frequency analysis has been around for centuries. In all that time, cryptographers have found several ways to improve on monoalphabetic ciphers, in an attempt to outwit the cryptanalysts. One method involves increasing the number of symbols used in the encryption, and encoding each letter in more than one way. This is known as *polyalphabetic* encryption. This can further be enhanced by introducing additional *dummy* symbols to the cipher. These do not represent anything at all, and will be deleted by the recipient of the message, but serve to confuse any cryptanalyst who intercepts the message. An example of a polyalphabetic code and message is:

a	b	c	d	e	f	g	h	i	j	k	l	m	n	o	p	q	r	s	t	u	v	w	x	y	z	dummy
Q	R	1	V	A	@	P	I	B	=	T	C	J	5	F	2	W	K	X	L	G	Y	M	H	E	N	D
^	U	$	Z	~	?	#	7	6	Ø	!	O	*	∞	π	+	&	8	3	%	S	−	×	9	£	÷	4

L^!AEπGD8+FXB%6F53Q∞Z28~+QK4A%π$F*JD~51AJ6X3DBF∞LI~3OA~4+A
KX×6O4CM^!AB5%76KL4£X69~QK%7IFS8D3

● The first unbreakable code

The ultimate problem for monoalphabetic ciphers is that one letter is encoded in the same way each time, which allows a frequency analyst a way in. Polyalphabetic ciphers and other techniques can help with this, but ultimately frequency analysis still gives cryptanalysts a foot in the door. There are other methods of enciphering for which frequency analysis is completely powerless. A famous example is the *one-time pad*.

What this amounts to is encoding each letter of the text with a new monoalphabetic cipher. It relies on having a text to act as a key. Suppose that the key begins: 'alphacentauri'.

If we wish to encode the plain text 'sleepersawake', the way to proceed is first to replace each letter of both the plain text and the key with the number that gives its alphabetical position, so 1 for A up to 26 for Z:

Key	a	l	p	h	a	c	e	n	t	a	u	r	i
	1	12	16	8	1	3	5	14	20	1	21	18	9

Plain text	s	l	e	e	p	e	r	s	a	w	a	k	e
	19	12	5	5	16	5	18	19	1	23	1	11	5

Now, to encipher the message we add the numbers in corresponding positions. If the result is 27 or higher, we subtract 26 from it. Then we replace the resulting numbers with the corresponding letters, to reveal the enciphered message.

Cipher text	20	24	21	13	17	8	23	7	21	24	22	3	14
	T	X	U	M	Q	H	W	G	U	X	V	C	N

How to break the unbreakable code

The one-time pad is literally an unbreakable code. The 13-letter text above could stand for any 13-letter plain text; there is no way to distinguish between them without having access to the key. Frequency analysis is of no use here, because one letter can be encoded in a different way each time it occurs. For example, the letter e occurs three times in the plain text above, and is encoded differently each time. The only way a one-time pad cipher can be broken is if the interceptor can guess, steal, or otherwise discover, the key. This is where the cipher's weakness is. Although on its own terms it is unbreakable, it requires a new key every time it is used (reusing the old key opens it up to cryptanalysis). Exchanging these numerous keys represents a security risk. For this reason, during wars, capturing enemy code books has often been a strategic priority. What these contain are keys for a series of one-time pads, or similar codes.

Codes in the 21st century

The codes that secure your email account are not one-time pads or monoalphabetic ciphers. They are a modern form of cipher that emerged in the 1970s, *public key cryptography*. As its name suggests, the major difference relates to the way the keys are distributed. For monoalphabetic ciphers, one-time pads, and the many variations on that theme, the key must be held by both the sender and receiver of information, and encryption and decryption are then symmetric operations.

Public key encryption functions in a totally new way. Here the key comes in two halves: public and private. Suppose someone wishes to send sensitive information to an internet-based company. The first thing to do is to access their public key, which is freely available to all. Then the message is encrypted using this public key, and sent to the company. However, decryption cannot be done using the public key alone; for the first time, the symmetry of encryption and decryption has been broken. Decryption requires the private key, and only the holder has this. In the most common forms of public key cryptography, the private key consists of two large prime numbers p and q, say. The public key is then the number $p \times q$. You might wonder whether it is not possible to deduce p and q from the number $p \times q$. The answer, probably, is that it is not. But this relies on one of the deepest questions in the subject (see *How to bring down the internet*).

30 How to avoid prison

- The prisoners' dilemma and game theory
- John Nash's equilibria
- Contracts and cooperative games
- Chess, checkers and artificial intelligence

Two corrupt financiers, Al and Bernie, are arrested for fraud. While they await trial, the police hold them in separate cells. The prosecuting lawyer, meanwhile, ponders a slight problem. There is currently only enough evidence to convict them of a relatively minor tax offence. For a successful prosecution on the more serious charge of fraud, he needs one of them to make a full confession.

After considering the situation, the lawyer visits Al in his cell, and offers him the following deal:

1. If you make a full confession and your partner refuses to do so, then with your testimony we can convict him of fraud. He will get ten years in prison, and you will walk free.
2. If you refuse to cooperate, but your partner confesses, then the opposite will happen. You will face ten years in prison, while your partner goes free.
3. If neither of you confesses, then we will have no choice but to accept the lesser charge of tax evasion. You will each get two years in prison.
4. If you both decide to confess, then you will each face eight years in prison.

While Al is left to think about it, the lawyer goes to visit Bernie, and makes exactly the same offer.

We will make the assumption that neither Al nor Bernie has any remorse for their crimes, or any ethical concerns about justice being done, or even any interest in the fate of their partner. Their sole interest is in minimizing their own jail-time. The question is: what strategy should the crooks follow?

Left alone in his cell, Al contemplates his options: 'For the two of us, the best option has got to be number 3. That has the least total jail-time, by a big margin. So maybe the best tactic is to keep schtum, and hope Bernie does the same. Wait a minute though . . . if he talks, then I'm looking at ten years. Come to think of it, *whatever* Bernie decides, I'm going to be better off confessing.'

So saying, Al decides to cooperate with the police. Bernie, in the other cell, reasons in exactly the same way, and comes to the same conclusion. They both confess, and the situation they end up with is option 4.

This situation is known as the *prisoners' dilemma*. The reason it is mathematically interesting is that it has a paradoxical flavour to it. Looking at the options above, and considering the prisoners as a pair, the best option is indeed number 3, by some distance. But the optimal strategy for each of the prisoners individually is to confess, and that leads to an eventual outcome of number 4, which is the worst of all the options for the pair of them.

● More than just a game

The *prisoners' dilemma* described above is a starting point for the subject of *game theory*. It has a long history, in terms of people applying mathematical reasoning to the study of traditional board games such as Chess and Go. It only recently developed into a fully blown mathematical discipline, however, with the work of Jon von Neumann, notably in his 1944 book with Oskar Morgenstern, *Theory of Games and Economic Behaviour*.

As this example illustrates, however, game theory has many applications far beyond games. Modern game theory is the science of strategy in all its forms; it is used today from economics to evolutionary biology, and is also relevant to many social sciences and even international relations.

The prisoners' dilemma

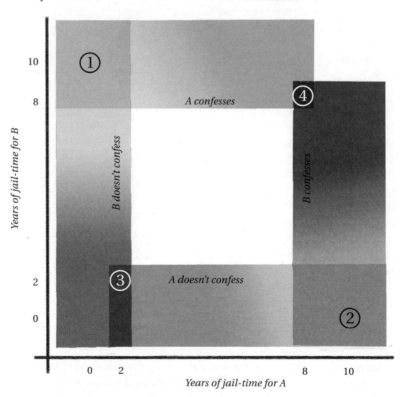

How to avoid prison

Beautiful minds and equilibria

One of game theory's most famous figures is John Nash. His doctoral thesis in Princeton in 1950 was a milestone in the subject. In 1994 Nash was awarded the Nobel Memorial Prize in economics. During his later life, Nash developed schizophrenia. He is the subject of the Hollywood biopic *A Beautiful Mind*.

A situation is called an *equilibrium* if no party has any incentive to change his strategy, even when made aware of the intentions of the other participants. In the prisoners' dilemma, option 4 is an equilibrium. Option 2 is not. If Al decides not to cooperate with the police, but is then informed that Bernie is planning to confess, he has an incentive to change his plans.

In other contexts there can be more than one equilibrium, but one of John Nash's outstanding achievements was to show that every game within a large class has at least one equilibrium state. He did this not just for games with two players, but those with three, four, or any conceivable number of players. What is more, he achieved this for so-called *mixed strategy games* in which players decide on their move probabilistically, rather than with certainty.

As the prisoners' dilemma shows, an equilibrium state need not represent any sort of best result. It simply means that no player can improve their own outcome unilaterally. In this sense, equilibrium states often correspond to traps into which the players may collectively fall. The only way to break out of them is through multilateral strategies, which entail a new ingredient: binding contracts.

How to keep a promise

Let's return to the prisoners' dilemma, but with a slight change. This time, let's suppose that, through a police oversight, Al and Bernie are left in the same cell for a few minutes, and take the opportunity to discuss strategy. Having thought about it, each realizes that, as things stand, they are each facing eight years in prison. If they work together, however, they can reduce this to two years. So they agree between them that neither of them will confess.

Once Al is back in his own cell, however, if he now believes that Bernie is not going to confess, he has an even greater temptation to confess and walk free. At the same time, he knows Bernie for the scoundrel he is, and does not trust him to keep his end of deal anyway. So Al decides to betray their agreement. In the other cell, of course, Bernie reaches the same conclusion. So they are back where they started.

The story would be different, though, if there was some way to enforce the agreement they had made. Suppose that (through serious police incompetence) Carlos, the local mafia boss, is allowed to visit both of them. In his presence, they each swear not to confess. Carlos makes it clear that, if either breaks their word, they will face a fate far worse than any amount of prison-time. This time, all goes smoothly, and they each end up with just two years each in prison.

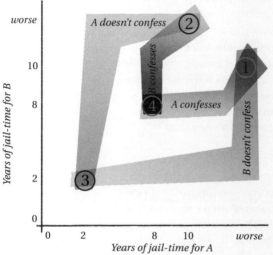

What this example shows is that the introduction of *binding contracts* fundamentally changes the rules of the game. Perhaps unexpectedly, it opens up previously unavailable opportunities that can be mutually beneficial. This insight is central to economics; indeed it has been of momentous importance in human history, where the trade of goods and services relies on enforceable contract law to punish swindlers and maintain customer trust. The theory of equilibria in the wider class of *cooperative games* (that is, games where contracts are allowed) is a topic of intense study and considerable economic importance.

The prisoners' dilemma with a binding contract

Contracts are not necessarily good things, of course. In the prisoners' dilemma, the police would be very keen to avoid the above scenario. (To avoid this type of situation, criminal fraternities often have a standing policy of non-cooperation with the authorities. This code of silence presents a serious obstacle to police investigating organized crime, as witnesses are often too frightened to break this code, well aware of what the punishment might be.) Similarly, if a group of competing businesses such as mobile phone providers all agree not to cut their prices below a certain level, this destroys the competition between them and forces consumers to pay more. This is why we have antitrust or competition law, to prevent such cartels emerging.

Computer games are changing the world

Game theory had its roots in the study of ancient board-games, such as Chess in India, Go in China, Nine Men's Morris in Rome, and Mancala in Africa. This original strand of the subject continues to generate a great deal of interest. This is particularly true since the dawn of the age of the computer. One of the fathers of modern computer science was Claude Shannon (see *How to talk to a computer*). In 1950, he wrote an influential paper 'Programming a computer for playing chess'. In it, he admitted that the problem was of 'no practical importance', but that it was deserving of consideration, since 'chess is generally considered to require "thinking" for skillful play; a solution of this problem will force us either to admit the possibility of a mechanized thinking or to further restrict our concept of "thinking".'

These were prophetic words. Since Shannon's work, game theory has become intimately bound up with the pursuit of artificial intelligence, and board-games such as chess have proved excellent test-beds for this research. So, when historians of the future look back on the evolving relationship between humans and machines, they may identify 11 May 1997 as a critical date. That was the day that IBM's machine Deep Blue completed its victory over the human world chess champion, Garry Kasparov. It defeated him 2 to 1 over six games, with three draws.

Part of Deep Blue's strength was its sheer computational power: it could analyze 200 million game positions per second. This certainly sounds a lot, but then again chess is a highly complex game. Shannon had estimated the number of different possible chess positions at around 10^{43}, that is, 1 with 43 zeroes after it. It would take Deep Blue incomparably longer than the lifetime of the universe to consider even a tiny fraction of these. So Deep Blue had to be able to reason in broad strategic terms as well as crunch numbers.

Checkers is a simpler game than chess, but with 5×10^{20} possible board positions it is still far too complex for direct computations of all possible moves to be possible. However, in 2007 a team led by the computer scientist Jonathan Schaeffer announced that they had completed an extraordinary analysis of this game. Using a network of dozens of computers running continuously since 1989, and applying state of the art techniques of artificial intelligence, Schaeffer and his colleagues were able to declare that 'Checkers is solved'. They had found a perfect strategy for playing, one that could never be beaten either by the greatest human player or any conceivable machine.

31 How to mislead a jury

- The prosecutor's fallacy
- The defence attorney's fallacy
- Thomas Bayes' conditional probability
- Rationality and Bayesian inference

Mathematics generally requires a great deal of time and effort. You cannot expect to grasp the nuances of complex geometry or the mysteries of the prime numbers without dedicating some time to studying them. The science of probability, however, is different. Everyone has some intuition for whether an event is likely or unlikely to happen, whether or not we have ever had formal classes in the subject. Indeed, it is likely that the essential tools needed to estimate likelihood are part of what make us human. Sometimes, however, this intuition goes haywire, producing answers that are wildly wrong.

The trouble with lawyers

Imagine that you are a prosecuting attorney in a bank-robbery case. There is only one piece of evidence against the defendant. A sample of DNA was found at the scene of the crime that closely matches that of the defendant. In fact, when the forensic scientist takes the stand, she says 'The genetic match here is 99.999%, meaning that only one person in 100,000 would produce a positive match to this sample'.

In your summing up of the case, you say 'Ladies and gentlemen of the jury, there may not be a large amount of evidence in this case. But what there is is compelling. We know that the perpetrator of this crime left the DNA sample at the scene. As you have heard, there is a 99.999% chance that this DNA is in fact that of the defendant. This is almost a certainty. On this basis, it is your duty to find him guilty.' You have just committed the *prosecutor's fallacy*.

Suppose that the city has a population of 5 million. Then the number of people there who would match the criminal's DNA is approximately 50. In the absence of any other evidence, there is no reason to think the defendant is more likely to be guilty than any of the others. In other words, the probability of guilt is not 99,999 in 100,000 but around 1 in 50, or 2%.

The term 'prosecutor's fallacy' originates from a 1987 paper by William Thomson and Edward Schumann, in which they investigated the use and abuse of statistics in criminal trials in the USA. They also performed experiments to see how readily people make this type of error. Their answer, in short, is very readily indeed. Of the people they interviewed, a majority were unable to spot the error, and went along with the fallacious reasoning. It even fooled at least one professional prosecuting attorney (so the fallacy is well-named). What is striking is that this is not a minor error. It is the difference between 99.999% and 2%; it could hardly be bigger.

To acquit the guilty or convict the innocent?

Thomson and Schumann also documented another common mistake, which they called the *defence attorney's fallacy*. The trial scenario outlined above is unrealistic in one important respect. If the forensic data really is the *only* evidence against the defendant, then (one should hope) the case would never make it to court. In any genuine trial, there should be other evidence as well: witness descriptions, circumstantial evidence, and so on.

Suppose that, in such a case, the defending attorney addresses the jury and carefully explains the erroneous reasoning of the prosecutor's fallacy. Then she says 'Therefore, ladies and gentlemen of the jury, I invite you to disregard the forensic evidence my colleague has presented. Although it seemed initially persuasive, as we have seen, it raises the probability of my client's guilt by a mere 2%. I remind you that if you are to reach a verdict of guilty, you are required to do so beyond reasonable doubt. Such insignificant evidence should have no place in your contemplations.'

'A jury consists of twelve persons chosen to decide who has the better lawyer.'

ROBERT FROST

The defence attorney is also guilty of a fallacy here. The result of 2% arrived at above relied on the assumption that, forensic evidence aside, the defendant was no more likely to be guilty than any of the other 4,999,999 inhabitants of the city. In other words, it assumes that there is no other evidence against him whatsoever. When there is other evidence, even if is weak and circumstantial, it will cut down the pool of potential suspects. For instance, suppose that eye-witnesses confirm that the perpetrator was male, between 5 feet 6 inches and 6 feet tall, between 15 and 35 years old, with black hair. This might be enough to reduce the pool of potential suspects from 5 million to around 200,000.

In the above case, the forensic evidence slashed the pool of potential suspects from 5 million to 50. Now it will reduce it from 200,000 to approximately 2. So we could legitimately argue that there is now around a 50% probability of guilt. With stronger non-forensic evidence, this figure will increase further, and could easily be damning.

When is an accurate test inaccurate?

The phenomenon of the prosecutor's fallacy has been long recognized in other contexts besides courtrooms. In fact it is widespread throughout the world. In the medical world, an important example is that of false positives.

Suppose I am randomly chosen to participate in a medical trial. Medical scientists have developed a new test for a disease; let's call it *bayesianitis*.

How to mislead a jury

The test is quite accurate: if I am suffering from the disease, there is a 99% chance that the test will correctly identify it, with just a 1% chance of it producing a false negative. If I am not suffering from it, there is a 95% chance that the test will come up clear, with a 5% chance of producing a false positive.

To my horror my result comes back positive, and I go home convinced that there is a 99% chance that I have the disease. Or should that be 95%? Either way, I have good cause to be very worried. Or don't I?

The correct answer depends on the prevalence of the disease among the general population. Let's suppose that it is rare, with around 1 in 10,000 people suffering from it. Starting with a population of a million people then, only around 100 will have the disease. When tested, around 99 of these will produce positive results. However, of the remaining 999,900 people who don't suffer from the disease, around 5% will also test positive, which comes out around 49,995 false positives. This already shows that the number of false positives hugely outweighs the number of true ones.

Out of a million people tested, this gives a total number of positive tests we would expect as 49,995 + 99 = 50,094. The probability that I am one of the 99 genuine positives, rather than one of the 49,995 false positives is then $\frac{99}{50,094}$, which is a small chance of around approximately 0.2%. So I really should not be too concerned.

Of course, it should be stressed that most medical tests do not fit this picture. For this reasoning to be valid, it is essential that I was *randomly* selected to take the test. If I start feeling sick and visit my doctor, who recognizes the symptoms of bayesianitis and recommends that I take the test, it would be completely inappropriate to apply the above line of thought. Indeed, to do so would precisely be to commit the defence attorney's fallacy. In this situation, the actual probability of my having the disease would depend on the prevalence of my symptoms, combined with a positive test, among the general population.

● The breakthrough of the Reverend Bayes

The same mathematical idea underlies the medical false positives, the lawyers' mistakes, and indeed a huge number of scenarios where probability theory gets used in the wider world. It is called *conditional probability*, and has its origins in the work of the Presbyterian minister the Reverend Thomas Bayes, who in the mid 18th century made some of the first investigations into this idea.

'Dúirt mé leat go raibh mé breoite' (I told you I was ill)

Inscription on Comedian
Spike Milligan's Gravestone

In the simplest examples of probability, we perform some experiment (typically tossing coins or rolling dice) and estimate the probability of a particular outcome. In the real world, however, things are interconnected, rather than stand-alone events. The interconnectedness of two different events is what conditional probability analyses.

Both the legal examples and the medical example are best considered in terms of conditional probability. In the medical test the two crucial events are A that I have the disease, and B that I test positive for the disease. What I am really interested in is the quantity denoted '$P(A|B)$', which represents the probability that I have the disease *given* that I test positive. However, all I know directly is $P(B|A)$, the probability that I test positive *given* that I have the disease, which is 0.99. These are the two central quantities, which it is all too easy to confuse. The central formula for understanding conditional probability is as follows:

$$P(A|B) = \frac{P(A\&B)}{P(B)}$$

That is to say, the probability of *A* given *B* is equal to the probability of *A* and *B* both occurring as a proportion of the probability of *B* occurring. In this case, the probability that *A* and *B* both occur is the probability that I have the disease ($\frac{1}{10000}$) multiplied by the probability that a sick person gets detected (0.99), which gives $P(A\&B) = \frac{1}{10000} \times 0.99 = 0.000099$.

On the other hand, what is $P(B)$, the probability that I test positive? Well there are two ways to test positive: through having the disease, and through not having it. That is to say $P(B) = P(B\&A) + P(B \& \text{not } A)$. We have just worked out $P(B\&A)$ as 0.000099. The probability of testing positive through not having it, that is, $P(B \& \text{not } A)$, is the probability that I do not have it (0.9999) multiplied by the probability of a healthy person testing positive (5% or 0.05). This comes out as $0.9999 \times 0.05 = 0.049995$. Adding together these two, we get $P(B) = 0.000099 + 0.049995 = 0.050094$. Now we can use the formula above, to get:

$$P(A|B) = \frac{0.000099}{0.050094} = 0.002 \text{ (to 3 decimal places)}$$

How to become fully rational

The problem of false positives and the lawyer's fallacies both come from our tendency to confuse the quantities $P(A|B)$ and $P(B|A)$, even when these are completely different. This is so widespread a phenomenon that some scientists suspect that it is an innate cognitive bias, an ingrained deficiency in human rationality. This makes conditional probability more than merely a useful mathematical technique. It is a crucial tool for understanding reality, and managing our inherent irrationality.

The strand of thinking known as *Bayesianism* seeks to apply conditional probability to all aspects of the world, from financial crashes to the perception of the risk of becoming a victim of crime. In all cases, to estimate the likelihood of an event *A* happening, we begin with an initial estimate $P(A)$ called a *Bayesian prior*. Now, to refine this estimate, we need to incorporate as much available data as possible. We formulate this as a second event *B*. Then we apply the formula above, to work out the conditional probability $P(A|B)$, known as the *Bayesian posterior*.

The more data that become available, the more we can update the probability. In this guise, probability itself takes on a new flavour. No longer is it a matter of the frequency of certain events (as it was with basic experiments with coins and dice). Now probability becomes a reflection of our individual knowledge of the state of the world.

32 How to slow time

- The speed of light
- Albert Einstein's special relativity
- The end of simultaneity
- The twins paradox

Many years from now, in deep space far from any stars or planets, are two spaceships. The SS Einstein is whizzing along at a constant speed of one million miles per hour, while the other, the SS Minkowski, is floating in space, completely stationary. On each of the ships is a scientist. The question is this: if the two scientists do not know which ship they are on, is there any experiment they can do to find out? If they analyze the motion of a ball bouncing on the floor, or examine the steam drifting up from a kettle, might this give them any clue whether their ship is stationary or travelling at high speed?

Since the work of Galileo Galilei and Isaac Newton in the 17th century, we have known that the answer is no. Life on board the two spacecraft is indistinguishable. There is no conceivable experiment that the scientists can perform that will tell them apart. As far as basic physics goes, there is no difference between travelling at constant speed, whatever that speed may be, and remaining still.

A philosophical conundrum follows from this, which troubled many thinkers of earlier ages. If there is no possible way to distinguish a moving spacecraft from one that is stationary, then is the distinction really meaningful at all? Perhaps there really is no true difference, except in relative terms.

This is indeed the viewpoint that modern physicists adopt. In the example above, it would be equally correct to say that the SS *Minkowski* is travelling at 1 million miles per hour while the SS *Einstein* is stationary. All that can be determined is the relative speed of one compared with the other. So we could validly view them both as travelling at 0.5 million miles per hour, in opposite directions.

It all depends on where you stand

This fundamental concept of *relativity* long predates Albert Einstein's famous work; it was first identified by Galileo Galilei in 1632. It seems alien to us on first meeting, because we Earth-bound humans are so used to having a fixed frame of reference by which to judge the motion of everything else. For most purposes, whether something is deemed stationary or moving is determined by its relationship to the ground. But this is a cosy illusion. In our corner of the universe nothing is stationary, not even in relative terms. The Earth is revolving on its own axis, while also orbiting the sun, which is

Motion viewed in three different frames of reference

in turn rotating around the centre of our galaxy. Perhaps the old concept of absolute space and absolute stationarity would not be much use to us anyway.

In any event, relativity does not apply to rotating objects; there are experiments that can distinguish between rooms that are rotating and those that are not, as anyone who has been on a merry-go-round will attest. It is only straight-line journeys, at constant speed (not accelerating), which are all considered equivalent.

How does one study physics once the idea of absolute stationarity has been abandoned, taking with it any concept of absolute location? The important concept is that of a *frame of reference*. Any room that is moving at a constant speed in a straight line has as good a claim to be stationary as any other. The differences between them are purely relative. So, at the start of a discussion, an astronaut might say 'for current purposes, I will take our ship, the SS *Einstein*, as stationary'. In so doing, she is specifying a frame of reference to work in, and all subsequent work will be performed relative to that. With this done, the speed of the SS *Minkowski* can be pinned down exactly. Of course she could well have started by taking the *Minkowski* as stationary, or indeed some other frame of reference according to which both ships were moving.

● Rømer's message from Jupiter

For hundreds of years scientists had been contemplating the nature of light. Does the light emitted by the sun reach us instantly, or does it travel at a finite speed as ordinary matter does, and as sound-waves do? Many of the great scientists of the past puzzled over this, but it was Ole Rømer in the 17th century who found the first evidence, by observing the sunlight reflected from Io, a moon of Jupiter. Rømer observed that when Io was further away from Earth the light took longer to reach us. The implication was that that light takes time to cover distance, and therefore travels at a finite speed rather than instantaneously. (A second question, whether it travels in waves or particles would be the key to the other great physical revolution of the 20th century, quantum mechanics. See *How to be alive and dead at the same time*.)

Although light travels at a finite speed, it is very fast (which is why the question had been so difficult to answer). With the development of more sophisticated equipment, that speed has been pinned down to 299,792,458 metres per second (around 670,616,629 miles per hour). To save writing this out, scientists generally refer to this speed as c, for short.

The strange arithmetic of light

Albert Einstein was surely the most famous scientist of the 20th century. But his celebrated work on the theory of relativity was not the product of a brilliant mind in isolation; it was also prompted by some remarkable and unexpected evidence about the nature of light.

If two balls are rolled towards each other, each travelling at 5 metres per second, then their *relative* speed – the combined speed at which they will collide – is 10 metres per second. So the same thing should be true for light. If two beams of light are fired at each other, each travelling with speed c, then we would expect the relative speed to be $2 \times c$. Similarly, if you are in a car travelling at 30 miles per hour, and another car overtakes you at 70 miles per hour, the second car seems to be travelling more slowly from your perspective than that of a pedestrian standing at the side of the road. It has a relative speed of 40 miles per hour compared with you. We should expect the same thing to be true of light: if you are travelling in the same direction as a beam of light, it should appear slower than if you are stationary.

This is what the old Galilean principle of relativity predicted, but in the early 20th century experimental evidence started to come in suggesting something which shocked the world of physics, namely that this line of reasoning fails, when light is involved. Its speed seemed to be absolute, rather than relative. However it was measured, no matter what the speed or direction of the person observing it, the speed of light always came out the same, as c. To account for this unexpected fact demanded a complete reconfiguration of the old ideas of space and time.

According to the new theory of *special relativity*, it is no longer true that all frames of reference travelling at constant speed are equivalent, only those that are travelling below the speed of light. Once the speed of light is reached, the old equivalence breaks down. In particular, there is no way for a human being or indeed any material object ever to catch up with a beam of light. No matter how fast you get, the beam of light always seems to be travelling at 299,792,458 metres per second faster than you.

'When a man sits with a pretty girl for an hour, it seems like a minute. But let him sit on a hot stove for a minute and it's longer than any hour. That's relativity.'

ALBERT EINSTEIN

● The end of time as we know it

The discovery of the absolute nature of the speed of light had dramatic consequences. If the motion of light is not relative, as all other motion is, this forces many other quantities that we had previously considered absolute to be relative after all. These include distance, mass, energy and, most controversially, time itself.

Imagine two different events occurring out in space, perhaps a balloon popping and a gun firing. If we want to compare the two, there are two obvious questions we might want to ask: whether or not they happen at the same time and whether or not they happen in the same place. The one thing we can determine for sure is when both of these are true: if the two events happen at the same place and at the same time. Beyond this, Galilean relativity had already dispensed with any absolute notion of being in the same place at different times: whether or not two events are in the same place depends on the frame of reference you select as your preferred notion of stationarity. Einstein's theory of special relativity did the same for the idea of two events happening at the same time in different places. *Simultaneity* was one of the first casualties of special relativity.

If it is possible for a physical object such as a human observer to travel from one event to the other and be present at both, then it makes sense to conceive of them as being 'at the same place and different times', since there is a frame of reference according to which this is true. This was true in Galilean relativity.

What is more surprising is that, if no material object can make this trip, and nor can a beam of light travel quickly enough to be present at both events, then we may legitimately describe the events as happening at the same time, in different places. If we prefer, however, we can conceive of them as happening at different times, in different places. This was a completely unexpected development: in all previous conceptions of the universe, events came in a strict order, irrespective of where they took place. This is no longer true in special relativity.

It was Einstein's teacher Hermann Minkowski who realized that it no longer made sense to consider time and space as separate entities, but as two aspects of a single 4-dimensional *spacetime*. Minkowski saw the potential of the new subject of *hyperbolic geometry* to describe the geometry of light (see *How to draw an impossible triangle*).

The twins with different ages

Let's imagine that in the year 3000, a pair of 30-year-old twins performs an experiment: Belinda sets off on a high-speed trip to Alpha Centauri, the nearest star to Earth after the Sun. She travels on board the SS *Lorentz*, which can travel at around 90% of the speed of light. Meanwhile Albert remains on Earth, living life with his family as he always has. He bids his sister farewell, expecting to see her again in ten years' time, and indeed in the year 3010 Belinda's rocket returns to Earth.

The twins are happily reunited but, when they compare notes, a strange story emerges: according to Belinda, her trip lasted only five years. The spacecraft's clock confirms this, and a doctor verifies that she has aged by exactly five years in the intervening time: she is now 35 years old and Albert is 40; the twins have been separated.

This *time dilation* is a consequence of the theory of special relativity. It is important to stress that, as far as Belinda was concerned, there was nothing unusual in her trip. Time ticked away in the ordinary way on board the spacecraft; it is just that time there and time on the Earth gradually drew apart as a consequence of the ship's high speed relative to the Earth. With an even faster rocket, the effect would be even more pronounced. If a rocket could be built that travels at 99% of the speed of light, time on the ship would run seven times faster than that on Earth. At 99.9%, it would run around 22 times faster. Of course a ship can never travel with speed c, but there is no theoretical obstacle to any speed below that, and therefore time may run as fast as you like, compared with that on Earth. Hendrik Lorentz analyzed the different rates at which time flows in different frames of reference.

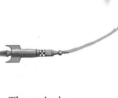

*The twins'
paradox*

Time dilation is an extraordinary idea, but it does work in practice. In 1971, Joseph Hafele and Richard Keating organized for highly accurate atomic clocks to be set on board commercial air-flights. As predicted by Einstein and Lorentz, the travelling clocks gained around 0.0000003 seconds compared with their Earth-bound counterparts.

33 How to win at roulette

- The St Petersburg paradox
- Averages – mean and expectation
- The law of large numbers
- The gambler's fallacy

There is a way to win at roulette every time. It works like this: first put £1 on red. If you win, that's the end: you have won, congratulations. If not, then next roll, put £2 on red. If you win this time, you walk away with £1 winnings in total. If you lose, then so far you are £3 down, and you put £4 on red, next roll. The system carries on like this, betting double the money each time. (Of course you do not have to bet on red every time; any sequence of red and black, or odds and evens, will do.)

Eventually, one of your bets will win. Suppose it takes six rolls to win. Then your losses up to that point will have been $1 + 2 + 4 + 8 + 16 = 31$, but on the sixth roll you bet £32 and win, and so end up with £1 profit.

This system is guaranteed to work, if followed exactly. But before you rush out to your nearest casino, let me add some important warnings:
1 There is no limit to the number of possible bets you might need to place in order to win. This is significant because the amount of money you need to bet on each successive roll increases, very fast. If you lose nine times in a row, then your next roll will be £1,024. If you lose 16 times in a row, then your next roll will be £131,072. By 20 rolls, you are betting millions of pounds. If you run out of money at any stage, that is the end, and you are left broke. The system has nothing to save you. So, the system assumes that you have infinitely deep pockets.
2 Although you are guaranteed to win, you are only guaranteed to win £1. Each time, you bet you are betting £1 more than all your previous losses. You could, of course, increase your profits by betting £100 more each time. This will mean you need even deeper pockets. By seven rolls you have to bet £12,800.
3 Many casinos have a maximum allowed bet on roulette. This essentially scuppers the system. If the maximum permitted bet is £10,000, and you need to place £16,384 for your fifteenth bet, then there is no way around it; you have lost.

● The paradox of St Petersburg
The system for winning roulette described above is interesting from a mathematical point of view. In the 18th century, Daniel Bernoulli considered the same system, in a slightly different context. Suppose you are offered to play a new game called St Petersburg. The rules are simple: you and I toss coins until we get a head. If we get a head immediately, you win £2. If we get a tail first, followed by a head, you win £4. If we get two tails followed by a

head, you win £8, three tails followed by a head wins you £16, and so on, with the prize doubling with every extra toss that is needed. Of course you are guaranteed to win something at least, so you will have no hesitation in playing. There is a catch, however: you have to pay to play. The question Bernoulli considered is: what is a reasonable price to play this game?

If the entry fee was £2 or less, then you are guaranteed to break even at least, so of course you would decide to play. If the entry fee was £3, then you might be prepared to accept a 50% chance of coming out £1 down, on the basis that you have a reasonable chance of winning more than that. What if the entry fee was £100? Unless you have money to burn, it is almost certain that you would decide against it.

How might a gambler decide whether a game is worth playing? The commonest method is to work out the average payout for the game, and compare this with the entry price. If the entry fee is less, then the game is worth playing; if it is more, then it is not.

● The infinitely profitable game

The simplest game of all is single coin toss. If I get a head, I win £10, if I get a tail I don't. What is the average payout here? Well, I have a chance of $\frac{1}{2}$ of winning £10, and $\frac{1}{2}$ of winning 0. To calculate the average, I multiply the various possible payouts by their probabilities and add them all together. In this case I get an average of $\frac{1}{2} \times £10 + \frac{1}{2} \times £0 = £5$. So if the entry price for single coin toss is less than £5, then I should play, and if it is more then I should not. If the entry fee is £4, and I play the game enough times, I can be confident of coming out ahead in the long run.

Something very strange happens when we apply this line of reasoning to Bernoulli's game of St Petersburg. Here you have a chance of $\frac{1}{2}$ of winning £2, a chance of $\frac{1}{4}$ of winning £4, $\frac{1}{8}$ of winning £8, $\frac{1}{16}$ of winning £16, and so on. Adding these together, we get $\frac{1}{2} \times 2 + \frac{1}{4} \times 4 + \frac{1}{8} \times 8 + \frac{1}{16} \times 16 + \cdots$, which is $1 + 1 + 1 + 1 + \cdots$. In other words, the 'average payout' of this game seems to be infinite.

This suggests that, however much the entry fee is, it always represents good value. Suppose that the entry fee is £5. After eight games you will have spent, $8 \times £5 = £40$. How much might you win? On average you could expect to win £2 four times, £4 twice, £8 once, and £16 (or more) once as well. That means a reasonable expected return is $4 \times £2 + 2 \times £4 + £8 + £16 = £40$, so you break even at this stage. With a few more games, you can expect to be in profit.

No matter what the entry fee, if you play enough games, you should expect to come out ahead eventually. If the entry fee is £22 per game, it will take you around a million games to come out on top (that is, around 2^{20}). Of course, as well as having several weeks spare to spend flipping coins, you must also have £22 million in cash to weather the losses in the mean time.

What do you mean, expectation?

The reason that mathematicians find the St Petersburg game interesting is that it shows the notion of 'average' breaking down in an unexpected manner. When the notion of average is applied to probabilities as in this example, mathematicians tend to call it the *expectation*. So the *expected payout* of single coin toss is £5. (This is a poor choice of terminology, as winning exactly £5 is one thing you can be certain will not occur. The expected payout of St Petersburg is infinite, which gives an even more misleading picture.)

When the average applies to a group of numbers, rather than probabilities, mathematicians prefer to call it the 'mean'. While the expectation is a more theoretical device considered by probability theorists and gamblers, the mean is common in daily life. To calculate the mean of a group of numbers, we add them all together and divide by how many numbers there were. For example, if the heights of the people in my family are 1.5 m, 1.8 m, 1.2 m and 1.7 m, then the mean height is $\frac{1.5 + 1.8 + 1.2 + 1.7}{4} = 1.55$ metres.

Two Bernoullis and two averages

So there are two important notions of average: the expectation, which comes from probability theory, and the common average, known as the mean. How are they related? The answer is supplied by one of the first great theorems of probability theory, *the law of large numbers*. This is the closest we get to the often-invoked urban myth, the 'law of averages'. The law of large numbers was proved by Daniel Bernoulli's uncle, Jacob Bernoulli, in 1713. (There are many other mathematical members of the Bernoulli family, just as there are many other important notions of average.)

Suppose we are flipping coins again. No money is changing hands; it is just for fun this time. I'm keeping score counting heads as 1 point and tails as 0. Of course the expected score is $\frac{1}{2} \times 0 + \frac{1}{2} \times 1 = \frac{1}{2}$, or 0.5. That is what probability theory tells us, but how does this relate to reality? If I toss the coin ten times, the mean score is the total number of points (that is, heads) divided by the total number of tosses, 10. Suppose I get six heads and four tails, then my average score is 0.6, not the 0.5 predicted by the theory.

Of course this is an entirely possible scenario, so does that mean that we can chuck the whole concept of expectation in the bin? Jacob Bernoulli's law of large numbers warns us not to be too hasty. What it says is that, the more times I toss the coin, the closer the mean score will get to the expectation. So if I toss the coin 100 times, I might get a mean score of 0.57, for a thousand times I might get 0.465, and over time the result will zero in on the expected score of 0.5.

The gambler's fallacy and fate

On a roulette wheel, there are 18 red grooves, 18 black grooves and 1 green groove, making a total of 37 grooves. The law of large numbers says that, over the long term, the average number of reds and blacks on roulette will be the same: each will appear around $\frac{18}{37}$ of the time, which is just under a half. Now suppose you are playing, and red comes up eight times in a row. To maintain the overall balance between red and black, surely a black must now be due?

No. This is the *gambler's fallacy*, the commonest of all misconceptions about probability. The probability of a black on the next spin is exactly the same as it was for the previous spin, and the spin before that. The law of large numbers makes absolutely no predictions whatsoever about the outcome of individual spins, only for the longer-term pattern of results.

So the law of large numbers does not offer the individual gambler a way to play the system. It does, however, explain why Las Vegas casinos are huge and lavish. The answer is that, on average, they win all the games. Individual gamblers might win individual spins of the roulette wheel, or come out ahead for a day, or a week. Some individual gamblers may even win over longer periods, through skill or exceptional luck. But, on average, the punters lose, and the casino wins.

In roulette, it is the green 37th groove, labelled 0, that is the big winner for the casino. When people are just betting red and black, the casino will take from the losers and pay out to the winners. So, on average, it will break even. But the law of large numbers guarantees that, on average, on every 37th spin, the green groove will appear, and both red and black will lose.

Once, a gambler asked Albert Einstein whether he knew a good strategy for playing roulette. The story goes that he paused for a moment, and then said that, yes, he knew a system that might possibly work. The gambler became very excited, and begged the great physicist to divulge his secret. Einstein replied 'steal the chips when the croupier is not looking'.

'Gambling: The sure way of getting nothing
for something.'

34 How to have beautiful children

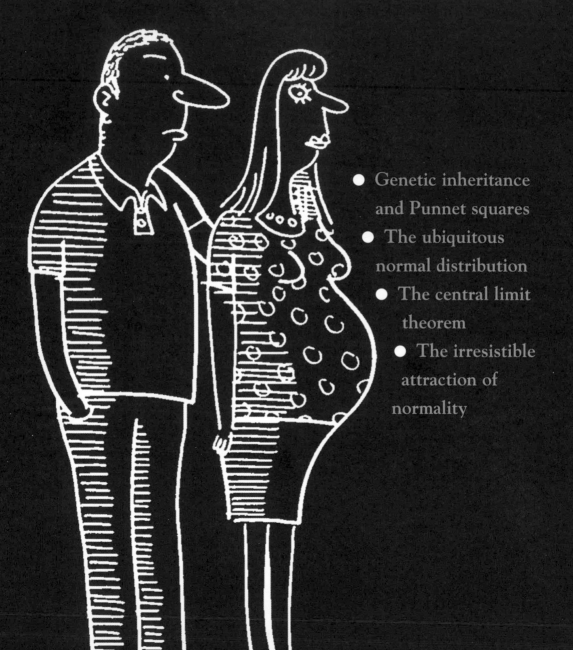

- Genetic inheritance and Punnet squares
- The ubiquitous normal distribution
- The central limit theorem
- The irresistible attraction of normality

Why are some people tall and others small? Why do some have straight black hair and others curly red? The answers to these and many other such questions are written in every cell of our body, in our genes – at least, unless we have had perms or plastic surgery. Our genes of course come from our biological parents. When their cells merge together to form us, their genes are divided up and mixed together to form ours. Hence we inherit many of their traits.

The story is not quite as simple as this though. It is not uncommon for two black-haired people to produce a blonde-haired child, for example. Many traits can jump generations, passing from grandparent to child, missing out the parents. Indeed, every few years, an otherwise healthy and normal human baby is born with a tail; this is a throwback to our ancestors of long ago. Such phenomena are examples of the unusual arithmetic of genes. We can best access this through the mathematics of *probability theory*.

The fundamental rules of genetics are the same as for random events such as tossing a coin, or rolling dice. If you toss a coin, the probability that it will land heads up is 50% (assuming the coin is not weighted and that you have not mastered the art of unfair flipping). Mathematicians generally prefer to express this as a fraction, $\frac{1}{2}$. This comes from the fact that there are two possible outcomes: heads and tails. The number of outcomes goes on the bottom of the fraction. However, we are interested in one of the outcomes, namely heads. The number of outcomes we class as *successes* goes on the top of the fraction.

Similarly, when I roll a die, the probability of getting a 6 is $\frac{1}{6}$. What is the probability of getting a 5 or a 6? In this case, there are now two successful outcomes, giving an answer of $\frac{2}{6}$ (which we can rewrite as $\frac{1}{3}$, around 33.3%).

The genetic coin toss

The same rules that apply to coins and dice also apply at a genetic level inside us (and indeed in all life on Earth). To take a simple example, suppose a type of dog comes in two varieties: black and white. The colour of the parent dogs seems to influence the colour of the offspring, but it is occasionally the case that two black dogs can produce a white puppy. On the other hand, it has never been observed that two white dogs produce a black puppy. What is going on here, at the genetic level?

Many simple attributes of an organism are determined by one gene. But this gene is composed of two smaller units, called *alleles*. One allele comes from

each parent. Suppose there are two possible alleles for the dogs' colour: B (for black) and w (for white). If the dog has a gene with two black alleles, BB, then it will be black. If it has two white alleles ww, then it will be white. The interesting question is what happens when it has one of each, Bw. This depends on which of the two alleles is *dominant*. In this case, the black allele is dominant (which is why we use a capital B but a lower case w). So the genotype Bw will produce a black dog.

Now, what happens during reproduction? Suppose the two parent dogs each have genotype Bw. The parents' alleles are each split up, and the puppy takes one allele from its mother, and one from its father. Each choice is made essentially at random. So there are four possibilities:

	Mother	
Father	B	w
B	BB	Bw
w	Bw	ww

One of these four combinations will be chosen during reproduction. In this case, what is the probability of the puppy being black?

We calculate it in exactly the same way as for the rolling dice or tossing coins. First we count the total number of possible outcomes, which in this case is four (as the table has four entries). Next we count the successes, meaning a black puppy. There are three of them: the single BB and the two Bw. This gives an answer of $\frac{3}{4}$, as the probability of the puppy being black. The same reasoning gives $\frac{1}{4}$ as the probability of a white puppy, coming from the single ww.

How to have beautiful children

If either of the parent dogs has genotype BB, then the puppies are sure to be black, since every puppy must have either genotype BB or Bw. Of course, this second possibility means that some of the next generation, the Bw dog's grandchildren, could still be white. If both parents are white, then certainly all the puppies must have genotype ww, and so also be white. However, what if one of the parents has genotype Bw, and the other ww?

Father	Mother	B	w
w		Bw	ww
w		Bw	ww

In this case the puppy has a $\frac{2}{4}$ (that is to say $\frac{1}{2}$) probability of being white.

These tables for understanding the arithmetic of inheritance are called Punnett squares after the geneticist Reginald Punnett, who first employed them in 1905. These squares encapsulate the process by which genes are divided up during reproduction. However, the situation is far more complicated than this suggests, as most aspects of our bodies are not just determined by one gene as in this example, but by a combination of several. Indeed, human DNA contains only around 25,000 different genes, less than twice the number of a fruit fly. Yet there are incomparably more factors than this that vary from human to human, even before taking environmental factors into account. It is the interaction between these genes that leads to the complexity and diversity of humanity.

For this reason, *gene expression* is a long-standing topic of study for biochemists. Individual traits such as height or hair colour are the result not of single genes, but of combinations of genes. What is more, individual genes are known to play roles in different and seemingly unrelated areas.

'I'm one of those people you hate because of genetics. It's the truth.'

BRAD PITT

We are all normal
The essential arithmetic of genetics is concisely encapsulated in the Punnett squares above. Yet when it comes to a trait such as height, people are not rigidly either short or tall; they come in a range of heights. There are the very rare cases of extreme shortness (0.56 metres is the unofficial world record at time of writing, claimed by Khagendra Thapa Magar of Nepal) and tallness (the American Robert Wadlow was 2.72 metres when he died in 1940). In between these any intermediate height is possible.

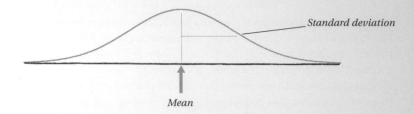

The normal distribution

Standard deviation

Mean

If we collect data of the heights of adult men and plot them on a graph, a very familiar shape appears. Most people know it as a *bell curve*, but to mathematicians it is the *normal distribution*. The normal distribution is one of the most important tools in data analysis, appearing in an extraordinary number of different contexts. Indeed, it is sometimes called the *Gaussian distribution* after the great Carl Friedrich Gauss who first applied it to astronomical data. The fact that human heights and the motions of planets should share a similar distribution is not some strange coincidence; it stems from a stunning mathematical fact at the heart of modern probability theory.

A normal distribution is specified by two numbers: its *mean*, which says where it is centred, and its *standard deviation*, which says how spread out it is. For the height of adult male humans, the mean in the USA is around 1.7 metres, with a standard deviation of roughly 0.1 metres. For females, the mean is 1.6 metres, and the standard deviation around 0.1 metres. (It is more illuminating to consider men and women separately, as the joint distribution is a more complex double-peaked curve.)

The irresistible attraction of normality

Of course, a person's environment does affect their height, especially during childhood. Diet and exercise are known to affect it, and indeed various techniques for poise and posture claim to be able to increase people's height by a few centimetres, even in adulthood. Nevertheless, biologists believe that the primary factor for determining height is genetic, at approximately 80%. How is it that the simple yes/no arithmetic of Punnett squares is replaced by the smooth range of variation described by the mathematicians' favourite, the normal distribution? These two pictures could hardly be more different.

Yet it is of no surprise to mathematicians that a morass of combining and overlapping Punnett squares should end up producing a normal distribution. One of the things for which the normal distribution is famed is its tendency to appear, even when least expected. The normal distribution is a *continuous* distribution: its set of possible outputs form a smooth range, not like the Punnett square, which has a *discrete* set of outcomes, like tossing

How to have beautiful children

a coin, or rolling a dice. In a discrete distribution, the possible outcomes are separated from each other. Yet, even when an experiment is discrete, the normal distribution can still apply. This is the result of one of the greatest theorems in probability theory, the *central limit theorem*.

It was discovered in 1733, by Abraham De Moivre, who realized that the normal distribution was excellent at modelling the number of heads in a sequence of coin tosses. Suppose De Moivre tossed 100 coins, and added up his total number of heads. The answer had to be somewhere between 0 and 100. If we plot the probabilities for all these on a graph, the result looks rather like the normal distribution. The two extreme answers are staggeringly unlikely (assuming the coin is fair, of course) and sit in the distribution's tail, and the most likely individual answer is 50, which sits in the centre, at the peak of the distribution. This is an example of the central limit theorem: it says that if you repeat almost any experiment, no matter how strange or abnormal, and add up the total score as you go along, the result gets ever closer to a normal distribution the more repetitions you perform. What is striking about this is that there is no requirement for the experiment to be anywhere close to normal. It could be a 50/50 coin toss, a Punnett square, or a much more exotic experiment with a hitherto unheard of distribution. It does not matter: almost all will converge on the normal distribution eventually.

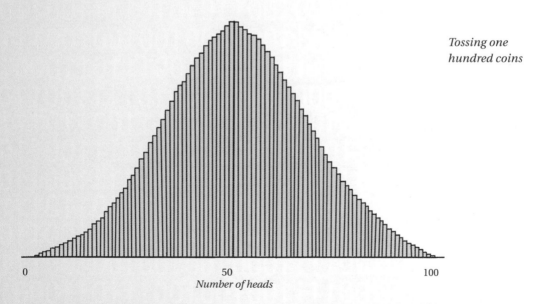

Tossing one hundred coins

0   50   100

Number of heads

35 How to talk to a computer

- How to do maths in Latin
- Zero and numerical notation
- Binary and other bases
- Claude Shannon's information theory

10111001011101111000100110101011110011011
11001011101111000100110101011110011011111
10111001011101111000100110101011110011011
10111001011101111000100110101011110011011
11001011101111000100110101011110011011111
10111001011101111000100110101011110011011
11000001101110100111100010010010010011011
00001101110100100111100100100111011000010
10111001011101111000100110101011110011011
11001011101111000100110101011110011011111
10111001011101111000100110101011110011011
10111001011101111000100110101011110011011
11001011101111000100110101011110011011111
10111001011101111000100110101011110011011
11000001101110100111100010010010010011011
00001101110100100111100100100111011000010
10111001011101111000100110101011110011011
11001011101111000100110101011110011011111

Why are there 60 seconds in a minute and 60 minutes in an hour? There is no reason from the natural world that says that it must be so; it is a cultural relic from the ancient Babylonian civilization of around 2000 BC. While we use a decimal system for representing numbers, one based around the number 10, the Babylonians largely employed a sexagesimal system, meaning that it is based on the number 60. There is nothing magical about the numbers 10 or 60; any number can serve as a base for representing numbers. Indeed, the base on which the modern world is built is not 10, but 2. This binary system is the language that computers speak.

How to do maths in Latin

Most of us learn how to read and write numbers at a young age. Once we have mastered it, we seldom pause to reflect on the system we use. We do not realize, therefore, what a beautiful and efficient construction it is, finely honed over the centuries.

It is a long way from the first system that people used to represent numbers, a tally. Here, to represent the number 12, you make 12 notches: ||||||||||||. The problem is obvious: you cannot recognize the number at a glance; you have to stop and count. Did I make a mistake and write 13 instead? The system is unreliable and slow to check. The second problem is that to represent large numbers is a nightmare. If there are 127 people in my village, I have to write this as:

||||||||||||||||||||||||||||||||||||||||||||||||||||||||||||||||||||||||||||||||||||||||||||||||||||||||||||||||||||||||||||||||||

This is profoundly unsatisfactory, and if I live in a larger settlement it is downright disastrous. The system can be improved by grouping the notches into clusters of ten (or indeed 60 or any other number) to make it more readable:

|||||||||| |||||||||| |||||||||| |||||||||| |||||||||| |||||||||| |||||||||| |||||||||| |||||||||| |||||||||| |||||||||| |||||||||| |||||||

Now we can count the clusters, instead of the individual notches. With this done, we might as well use a different symbol to represent a cluster of ten, such as X. So we get: XXXXXXXXXXXX|||||||

Now we have the same problem again: we have to count the Xs. So we might introduce another new symbol, C, to represent a cluster of ten Xs. To cut down that string of seven notches at the end, we can introduce another symbol V to represent five. Then the number becomes: CXXVII

This is essentially how Roman numerals developed, as shorthand for a tally, using the following symbols:

Symbol	Meaning	Symbol	Meaning
I	1	D	500
V	5	M	1,000
X	10	$\overline{V}$	5,000
L	50	$\overline{X}$	10,000
C	100	$\overline{C}$	100,000

The symbols are always arranged in decreasing order, as in MDCLXVI (1,666). (Much later, medieval writers also introduced a 'subtraction rule' for small digits preceding larger ones, meaning that nine could be represented as IX instead of VIIII.)

Zero changes the game

The system of Hindu–Arabic numerals that we use today has many advantages over Roman numerals. To start with, it has no upper limit. Roman numerals are satisfactory for representing numbers up to a million, but in the modern world we often need much bigger numbers. All we can do then is invent more symbols to stick on the front. (Indeed, this is exactly the situation that Archimedes found himself in when he set out to estimate the number of grains of sand that would be needed to fill the universe.)

Our modern system is extremely economical in symbols, and only ever requires ten symbols, no matter how large a number we are writing: 0, 1, 2, 3, 4, 5, 6, 7, 8, 9. Once we get to ten, we start to recycle these symbols, in a subtle way. Instead of simply counting the units, we introduce a new column to count the number of tens that fit into it. For still larger numbers we add another column representing the hundreds. Why is a hundred the crucial threshold? The answer is that it is 10×10 (or 10^2 for short). Then the fourth column appears once we reach a thousand, that is, ten hundreds, or $10 \times 10 \times 10$ (or 10^3).

This type of system is known as *place-value notation* as the position that the symbol occupies carries just as much information as the symbol itself. The symbol '6' might represent the number six, but it might equally represent

How to talk to a computer

sixty or six hundred thousand, or even six hundredths. It depends on where the symbol is positioned. In the numbers 6, 600,000 and 0.06 the symbol '6' carries different meanings. As this shows, the critical moment in the development of place-value notation was the introduction of a symbol for zero, which happened in seventh-century India.

How to be a mathematical polyglot

There is nothing special about the number ten. That this is the base of our number system is a consequence of biology rather than mathematics: we have ten fingers, on which we first learned to count. Any other number would serve equally well. For example, we could work in base seven. So instead of columns representing units, tens, hundreds and thousands, we would have columns representing units, sevens, forty-nines and three-hundred-and-forty-threes. (This sounds awkward because our language, as well as our notation, is decimal.)

To translate a number such as 139 from decimal into base seven, the first question to ask is: what is the largest power of seven that will fit into it? Seven can fit, as can 49, but 343 cannot. So the first column will represent the number of times 49 fits into our number, which is 2. How much of the number does this account for? Well, two 49s are 98, which leaves 41 left over. Seven fits into this 5 times leaving a remainder of 6. Putting all this together, we get an answer of 256.

It is important to realize that what is happening here is quite superficial. We are only talking about different ways to write numbers down. The number itself remains completely unchanged throughout this procedure, just as an animal is indifferent to whether it is described in English as 'mouse', in French as '*souris*', or in Japanese as '*nezumi*'.

'There are 10 types of people in the world: those who understand binary and those who don't.'

ANONYMOUS

The power of two

Any whole number larger than one can act as a base for representing numbers. While humans have largely settled on ten through accident of biology, computers have other priorities. For machines, the most convenient system is base two, known as *binary*. If we start counting in binary, it goes:

0, 1, 10, 11, 100, 101, 110, 111, 1000, 1001, 1010, 1011, 1100, 1101, 1110, 1111, 10000

Only two symbols are ever required: 0 and 1. This is extremely useful, as they can be stored mechanically or electronically as the 'off' and 'on' settings of a switch. Sequences of switches can then be used to store larger numbers. So if a sequence of five switches reads 'on on on off on' this represents the binary number '11101'. To translate this into decimals, the first column represents sixteens, the second eights, then fours, twos, and finally units. So these switches represent the number $1 \times 16 + 1 \times 8 + 1 \times 4 + 0 \times 2 + 1 \times 1 = 29$, in decimal notation.

Among many other achievements, Gottfried Leibniz was the first person to see the potential of binary for numerical expression, laying the foundation in his 1701 work 'Essay on a new science of numbers'. Leibniz set about translating reams of numbers from decimal into binary. Later in his life he even came to view binary as some sort of mystic language of creation, a dialogue between 0 representing the eternal void, and 1 representing the irreducible unity of God. In the tension between the two, the physical universe blossoms into existence.

Humans also use binary to communicate with each other; it is the basis of Morse code. In Morse code, each letter of the alphabet is translated into a system of dots and dashes. If we take a dot as representing 0 and a dash as 1, this amounts to encoding every letter as a number in binary.

Computers learn to talk

With the dawn of the digital computer, hundreds of years after Leibniz had the idea, binary established itself as the natural language of communication for machines. It is not just numbers that are encoded in binary, but letters, pictures, music, videos, spreadsheets, webpages, and the enormous range of other objects that modern computers are equipped to handle. Ultimately, all of these are reducible to long sequences of 0s and 1s.

Of course we seldom see these strings of binary digits, or *bits* as they are called. Not even those of us who build or program computers are expected to decipher strings of 0s and 1s that are thousands of bits long. This truly is the private language of the computers. When a human needs to interface with it, the machines generally do us the courtesy of translating the message into a more digestible language.

Since the birth of the internet, computers no longer spend their lives in isolation, but are in constant communication with each other. This means they need to exchange data. At this point, it becomes important to encode information efficiently. Just as humans found better ways to represent numbers than simple tallies, so computers needed to express information in binary strings in as concise a manner as possible.

In 1948, Claude Shannon published a paper 'A mathematical theory of communication', which inaugurated the subject of *information theory*. In it, Shannon analyzed the limits of information communication, in terms of binary strings being transmitted along a channel. Shannon considered both perfect channels and noisy channels in which data may become lost or corrupted. In this case, an additional challenge is to find *error-correcting codes* that allow the data to be reconstructed on reception. One technique involves encoding Latin squares into the binary strings (see *How to excel at Sudoku*).

How to measure information

To a human, one long string of 0s and 1s will look pretty much like another. However, this is simply because we do not naturally speak the language. In fact there is a huge variety, most importantly in terms of the amount of *information* that a string contains. In extreme cases, we can see this. For example, it is clear that a sequence that consists of one million 0s contains little information, despite its great length.

What computer scientists needed was a way to quantify the essential information content of a string. The critical insight is that this string can be radically *compressed*. In fact I did compress it in the paragraph above. Instead of writing out the whole sequence (which would require a book to itself, the dullest ever written), I wrote 'a sequence that consists of one million 0s', which perfectly describes it in eight words of English.

Compression is a vital tool in computer science. On the practical side, it prevents your computer's memory being clogged up with huge amounts of redundant data. On the theoretical side, compression also provides a way to measure information exactly. If we start with a string of bits, the question is: what is the shortest sequence we can use to describe this string exactly? That is to say, what is the minimum length to which it can be compressed? Shannon called this number the *entropy* of the string, and it exactly quantifies the amount of irreducible information it contains. The strings of bits that are richest in information are those that are incompressible: they cannot be shortened. This means that each bit must be totally unpredictable from what has gone before; there can be no pattern within the sequence. Apart from anything else, this implies that the string must be *normal* (see *How to square a circle*). It is one of the paradoxes of modern logic that the sequences that encode the greatest possible amount of information are also those that are truly *random*, and could derive from something as meaningless as the tossing of a coin.

'It is a very sad thing that nowadays there is so little useless information.'

Oscar Wilde

Glossary

Algorithm
A list of instructions that tell a machine how to accomplish a particular task, known outside mathematics as a *computer program*.

Benford's law
A rule that describes the relative frequency of the first digits in a set of data. It predicts that the numbers 1 to 9 should not feature equally often as first digits. Though not inviolable, experience shows that Benford's law is often satisfied in practice.

Binary
A way of writing numbers that is based on the number two instead of the more familiar ten. So the number written in decimals as '9' appears in binary as '1001'. Binary is the natural language of computers.

Calculus
The science of rates of change. Given a process, the challenge is to find its rate of change. The algebraic rules were eventually discovered by Isaac Newton and Gottfried Leibniz.

Cellular automaton
A theoretical device comprising an array of cells that change colour according to fixed rules concerning the colours of neighbouring cells. Despite their simplicity, some cellular automata are capable of great feats of computation.

Chaos theory
The study of feedback processes, characterised by an extreme sensitivity to the exact starting conditions – the so-called *butterfly effect*.

Classification of finite simple groups
An enormous theorem, completed in 2004, which lists all the possible *finite simple groups*, the basic building blocks of symmetry. The theorem describes 18 families and 26 individual sporadic groups, which together cover all possibilities. The largest of the sporadic groups is the *monster*.

Combinations and permutations
The possible choices when objects are picked from a collection. Starting with 5 objects, the number of ways of choosing 2 is the number of *combinations* of 2 from 5, which comes out as 10. If we also care about the order in which they are chosen, this is the number of *permutations* of 2 from 5, which is 20.

Complex numbers
A 2-dimensional number system built around a new so-called imaginary number i, defined by $i \times i = -1$. The complex numbers form the ultimate number system, in that they satisfy the fundamental theorem of algebra.

Conditional probability
The chance of one event, given another. When rolling a fair die, the probability of getting a 6 is $\frac{1}{6}$, but the *conditional probability* of getting a six *given* that the result is an even number, is $\frac{1}{3}$.

Continuum
The second level of infinity, after the familiar *countable* level of the numbers 1, 2, 3, 4, . . . Georg Cantor in 1874 proved that the continuum is larger than the countable level, and that above it lie even higher levels.

Cryptography
The science of making codes, the opposite of which is cryptanalysis: the science of breaking codes.

Euler's formula
Discovered by Leonhard Euler, an equation of exquisite beauty, which connects the five most important numbers in mathematics: $e^{i\pi} + 1 = 0$.

Fermat's last theorem

The formula $a^n + b^n = c^n$ is obtained by replacing the squares in Pythagoras' theorem with higher powers. Pierre de Fermat believed that when the value of n is 3 or more this equation is not obeyed by any triple of whole numbers (a, b, c). It was proved by Andrew Wiles in 1995.

Fibonacci sequence

1, 1, 2, 3, 5, 8, 13, 21, 34, . . . A sequence of numbers defined by the fact that the next term is always the sum of the preceding two. Discovered in the 13th century by Leonardo Fibonacci, it features in several places in the natural world and is closely related to the golden section.

Finite simple group

A finite group that cannot be broken up into smaller groups (somewhat analogous to a prime number).

Four colour theorem

The fact that any map can be coloured with four colours so that no two neighbouring countries share the same colour. It was conjectured in 1852 by Francis Guthrie, and proved by in 1976 by Appel and Haken with a hugely long computer-assisted argument.

Fractal

A geometric object that looks the same however far you zoom in. Fractals are renowned for their intricate beauty, and more surprisingly are useful for modelling aspects of the physical world, including coastlines and commodity-markets.

Fundamental theorem of algebra

The assertion that the system of complex numbers solves every equation we could hope for. Technically speaking, every polynomial equation that can be built from complex numbers can also be solved using complex numbers.

Game theory

The science of strategy, now a large area of mathematics that grew from reasoning about board games such as Chess.

Gödel's incompleteness theorem

Kurt Gödel's seminal result of 1931 proved that no set of rules that could conceivably be written down would ever be enough to underlie ordinary arithmetic.

Golden section

The number $\phi = \frac{1 + \sqrt{5}}{2}$, around 1.62. It is defined to allow a line to be divided into two parts, where the ratio of the whole line to the longer part is equal to ϕ, as is that of the longer to the shorter. It is related to many pieces of mathematics, notably the Fibonacci sequence.

Group

An abstraction of the notion of addition. A group is a collection of objects that can be combined together satisfying three laws: that there is one special *identity* object that leaves every other unchanged (in ordinary addition this is 0, because $0 + 5 = 5$ for example); that every object has an inverse that cancels it out (the inverse of 5 is -5 because $-5 + 5 = 0$); and finally a technical condition known as *associativity*, which says that the way objects are bracketed doesn't affect the outcome: for example, $1 + (2 + 3) = (1 + 2) + 3$. Groups are widespread throughout mathematics. Some are infinite, such as the group of fractions under multiplication, and others finite, such as the collection of symmetries of a cube.

Hyperbolic geometry

Discovered in the 19th century, a radical new system of geometry that disobeys the parallel postulate (one of the fundamental laws underlying the ordinary system of Euclidean geometry).

Information theory

Begun by Claude Shannon in 1948, the study of how information can be encoded in strings of

0s and 1s, and then transferred efficiently, even along noisy channels. Although it was developed much earlier, information theory underlies the modern internet.

Irrational number
A number that cannot be written exactly as a fraction of two whole numbers. Examples include $\sqrt{2}$ and π.

Kepler's conjecture
A claim about the most efficient way to pack together spheres to use up the least space. Johannes Kepler claimed that the best method is to lay down one layer in a hexagonal pattern, with each sphere touching six others, and then lay a similar layer on top, sitting as low as possible, and so on. Although this arrangement is used by greengrocers around the world, it was not proved optimal until 1998, by Thomas Hales.

Knot invariant
An object that can be associated to any knotted loop, in such a way that if two knots are actually one in disguise, then they will produce the same invariant. A famous example is the Jones *polynomial* of 1984, and the search is continuing for invariants that are better at distinguishing ever wider classes of knots.

Latin square
A square grid in which each number appears once in each row, and once in each column. These are used in several areas of mathematics, including the design of error-correcting codes, and appear in newspapers daily as Sudoku.

Liar paradox
The classic logical paradox, due to Eubulides in the 4th century BC: 'This is a lie'.

Navier–Stokes equations
The fundamental equations that describe how a viscous fluid flows. Remarkably, no mathematical solutions are known.

Normal distribution
The commonest way in which data is spread, known outside mathematics as a bell curve. The *central limit theorem* predicts that many experiments, if repeated often enough, will result in normal distributions.

Parallel postulate
One of Euclid's fundamental axioms of geometry, stating that for any straight line, and any point not on it, there is exactly one new straight line that passes through the point, running parallel to the original line. For many years it was unknown whether the parallel postulate followed from Euclid's other axioms. But with the discovery of non-Euclidean hyperbolic geometry, it was finally established that it does not.

Perfect cuboid
A cuboid (i.e. a stretched or squashed cube) where all the sides have lengths given by whole numbers, as do the face diagonals and the body diagonals. It is unknown whether such an object can exist. Dropping the condition on the body diagonal gives an *Euler brick*, of which examples are known.

π
The number you get when you divide a circle's circumference (the distance around the outside) by its diameter (the distance across the middle). It is approximately 3.142, but its decimal representation carries on forever without repetition, as it is an irrational number.

Platonic solids
The five most symmetrical shapes that exist in three dimensions. They are the tetrahedron, cube, octahedron, dodecahedron, and icosahedron. Their classification has been known since the ancient Greeks.

P=NP?
The major question in theoretical computer science, which asks whether every task that can be checked quickly can in fact be completed quickly.

Poincaré conjecture

A statement about the possible 3-dimensional shapes that contain no holes. In the early 20th century, Henri Poincaré claimed that the only such shape is the 3D sphere (the 3-dimensional equivalent of the ordinary 2-dimensional sphere). This was eventually proved in 2002 by Grigori Perelman, a milestone in contemporary topology.

Prime number

A whole number, such as 2, 3, 5, or 7, which can only be divided by itself and 1. Primes are the building blocks of all other numbers, but a great deal is still unknown about them.

Prime number theorem

An estimate for the number of prime numbers up to a given threshold. It was conjectured by Carl-Friedrich Gauss in 1792, and proved by Jacques Hadamard and Charles de la Vallée Poussin in 1896.

Pythagoras' theorem

A foundational theorem of geometry of right-angled triangles. It says if the lengths of the sides are a, b, and c (with c the longest) then $a^2 + b^2 = c^2$.

Quantum mechanics

A physical theory, according to which matter consists of neither particles nor waves but has properties of both. It has many surprising consequences, including the paradox of Schrödinger's cat, but is well supported by experimental evidence.

Ramsey's theorem

A seminal fact that allows highly ordered substructures to be found even in very disordered situations. Originally discovered in relation to the dinner party problem, now in use throughout mathematics.

Riemann hypothesis

Perhaps the greatest unsolved question in mathematics; in 1859, Berhard Riemann described in detail how he thought the primes should be scattered among the whole numbers. His claim remains unproved today.

Series

A list of numbers that are added up as you go along. Some such as $1 + \frac{1}{2} + \frac{1}{4} + \frac{1}{8} + \cdots$ *converge*, that is get closer and closer to a fixed number (in this case 2). Others such as $1 + 1 + 1 + 1 + \cdots$ *diverge*, that is get bigger and bigger without limit.

Special relativity

A theory of physics according to which all speeds below that of light are indistinguishable, but this equivalence breaks down at the speed of light. It has many unexpected consequences, including the fact that time runs at different rates for observers travelling at different speeds.

Symmetry

A way of transforming a geometrical object, such as a square, which results in it looking the same. Examples include rotation and reflection.

Topology

An approach to geometry where two shapes are considered the same if one can be stretched into the form of the other.

Transcendental number

A number such as π, which can never result in a whole number, no matter how many times you multiply it by itself and add up the results. This is a stronger condition than being irrational.

Wallpaper group

One of 17 abstract designs, which together list all the possibilities for a repetitive pattern. This classification was proved in 1891 by Evgraf Fedorov.

Weaire–Phelan foam

A method of dividing 3-dimensional space into cells of equal size, using less material than any other known foam. It was discovered in 1993, and contradicted Lord Kelvin's conjecture about the optimal way to solve this problem.

Index

Figures in italics indicate captions.

Acknowledgements

Quercus Publishing Plc
21 Bloomsbury Square
London
WC1A 2NS

First published in 2011

A catalogue record of this book is available from the British Library

ISBN: 978 1 84916 480 1

Cartoons by Mike Mosedale
Illustrations and diagrams by Patrick Nugent,
except p.68/70 by Bill Donohoe; 124, 126, 127 by Pikaia
p. 44/47 © Science Photolibrary/Scott Camazine

Editorial and design management by BCS Publishing Limited, Oxford
Typeset by Lapiz Digital, India

Printed and bound in China

10 9 8 7 6 5 4 3 2 1